INSTRUMENTATION
Workbook

FIFTH EDITION

atp AMERICAN TECHNICAL PUBLISHERS
ORLAND PARK, ILLINOIS 60467-5756

Thomas A. Weedon

Instrumentation Workbook contains procedures commonly practiced in industry and the trade. Specific procedures vary with each task and must be performed by a qualified person. For maximum safety, always refer to specific manufacturer recommendations, insurance regulations, specific job site and plant procedures, applicable federal, state, and local regulations, and any authority having jurisdiction. The material contained herein is intended to be an educational resource for the user. American Technical Publishers, Inc. assumes no responsibility or liability in connection with this material or its use by any individual or organization.

© 2010 by American Technical Publishers, Inc.
All rights reserved

5 6 7 8 9 – 10 – 9 8 7 6 5 4 3 2

Printed in the United States of America

ISBN: 978-0-8269-3431-4

This book is printed on 30% recycled paper.

INSTRUMENTATION WORKBOOK

Contents

Section 1—Introduction to Instrumentation

Section 2—Temperature Measurement

Section 3—Pressure Measurement

Section 4—Level Measurement

Section 10—Final Elements

INSTRUMENTATION WORKBOOK

Introduction

Instrumentation Workbook is designed to reinforce the concepts, provide applications, and test for the comprehension of the material presented in *Instrumentation*, 5th Edition. The workbook closely parallels the organization of the textbook. Each section is divided into chapters with questions and activities that review and supplement the concepts presented in that chapter of the textbook.

Review Questions

The workbook contains 50 sets of Review Questions. Review Questions consist of multiple choice, true-false, and/or completion questions that are based on the text and art in the corresponding chapter and section of *Instrumentation*.

Activities

The workbook contains 128 Activities developed from the 12 sections in the textbook. Activities provide opportunities to apply the concepts and theory of each major section in the textbook to practical problems. See the *Instrumentation Workbook* Table of Contents for a complete listing of Activities.

Integrated Activities

The workbook contains 11 Integrated Activities. The Integrated Activities provide the opportunity for the learner to combine ideas and concepts from several sections of the textbook into one activity. The Integrated Activities help develop the higher-level analysis and problem-solving skills required in today's workplace.

Appendix

The Appendix contains charts and tables for use with the Activities. See the Appendix Table of Contents for a complete listing of the charts and tables in the Appendix.

Related Information

Information presented in *Instrumentation*, 5th Edition, and *Instrumentation Workbook* addresses common instrumentation topics. Additional information related to instrumentation is available in other American Tech products. To obtain information about these products, visit the American Tech web site at www.go2atp.com.

SECTION
1 INTRODUCTION TO INSTRUMENTATION

chapter
1

Instrumentation Overview

REVIEW
QUESTIONS

Name _____   **Date** _____

T   F   **1.** Process control instrumentation is the technology of using instruments to measure and control manufacturing, conversion, or treating processes to create the desired properties of materials.

_____   **2.** A significant trend in instrumentation is the replacement of simple, repairable ___ devices by sophisticated electronic systems.

T   F   **3.** Continuing education about process control instrumentation is not necessary.

_____   **4.** ___ maintenance technicians are becoming more popular in industry because they offer flexibility and cost-effectiveness.
         A. Specialty
         B. Multiskilled
         C. Untrained
         D. Inexperienced

T   F   **5.** A boiler operator is typically responsible for control systems that measure temperature, pressure, flow, and other variables.

T   F   **6.** HVAC technicians are often required to use instruments to take measurements in the field to troubleshoot problems.

T   F   **7.** A technician can stop learning and coast through a job after initial training.

T   F   **8.** A common source of training on instruments is from the instrument manufacturers.

T   F   **9.** Short courses to help technicians learn new skills are often available at trade shows.

T   F   **10.** There are many trade union training programs that help technicians gain and develop the skills needed to be successful.

Instrumentation Overview
ACTIVITIES

Name _____   Date _____

Activity 1-1—Fractions and Decimals

There are many applications that require an understanding of fractions, decimals, and percentages. Technicians need to understand the differences and how to convert from one to another.

Fractions are a part of a whole number. Fractions are represented by a whole number numerator above a whole number denominator. The fraction may be laid out as a two-line fraction separated by a horizontal line, or as a one-line fraction separated by a slanted line. For example, $\frac{1}{2}$ is a fraction that represents the same amount as ½ or 1/2.

Decimal numbers are represented as a series of numbers with a decimal point. For example, 0.5 and 0.667 are decimal numbers. Percentages are a number representing the number of parts of 100. For example, 50% represents 50 parts per hundred.

Convert the following fractions into decimal numbers.

_____ 1. ¼

_____ 2. ⅞

_____ 3. ⁹⁄₁₆

_____ 4. ⁷⁄₁₂

_____ 5. ¹¹⁄₂₀

Convert the following fractions into percentages.

_____ 6. ¼

_____ 7. ⅞

_____ 8. ⁹⁄₁₆

_____ 9. ⁷⁄₁₂

_____ 10. ¹¹⁄₂₀

Convert the following decimal numbers into fractions.

_____ 11. 0.23

_____ 12. 0.47

_____ 13. 0.83

_____ 14. 0.375

_____ 15. 0.625

Convert the following decimal numbers into percentages.

_____ 16. 0.23

_____ 17. 0.47

_____ 18. 0.83

_____ 19. 0.375

_____ 20. 0.625

Convert the following percentages into decimal numbers.

_____ 21. 13%

_____ 22. 3%

_____ 23. 67%

_____ 24. 35.5%

_____ 25. 183.2%

Name _____   Date _____

Activity 1-2—Calculation Order and Rearranging Equations

The order in which mathematical operations are done can have a major impact on the results. Operations within parentheses are done first. If parentheses are nested, start from the innermost set and work outward. After parentheses, the priority of operations is as follows:

1. Powers and roots

2. Multiplication and division

3. Addition and subtraction

Perform the following calculations.

_____   **1.** $[(2.9)^2 \times (30 - 3)^{1/3} + 1.03]^2$

_____   **2.** $[(5.9)^{1/2} - 2.1]^2 - [(8.0)^{1/3} - 1.1]^2$

_____   **3.** $\{(2.3 - 1.1)^3 \times [0.31 \times (5.7 - 2.6)]^2 \div (8.7 + 0.7)^{1/2}\}^{1/2}$

There are many times when it is necessary to rearrange an equation to solve for an unknown value. Rearrange the following equations to solve for the unknown term given.

_____   **4.** Ohm's law, $E = I \times R$. Rearrange to solve for R.

_____   **5.** Ohm's law, $E = I \times R$. Rearrange to solve for I.

_____   **6.** General equation of flow, $F = C \times \sqrt{\Delta P}$. Rearrange to solve for C.

_____   **7.** General equation of flow, $F = C \times \sqrt{\Delta P}$. Rearrange to solve for ΔP.

_____   **8.** Force and pressure, $F = P \times A$. Rearrange to solve for P.

_____   **9.** Force and pressure, $F = P \times A$. Rearrange to solve for A.

_____   **10.** Hydrostatic pressure, $P = H \times \rho$. Rearrange to solve for ρ.

_____   **11.** Hydrostatic pressure, $P = H \times \rho$. Rearrange to solve for H.

4

Name _____   **Date** _____

Activity 1-3—Scientific Notation

Many numbers used in industrial applications are very small or very large. These numbers may contain many leading or trailing zeroes. These numbers are hard to read because of the number of zeroes. Scientific notation is a method of using powers of 10 combined with decimal numbers to simplify the presentation of these large or small numbers.

An exponent represents the number of times a number is multiplied by itself. For example, 10^2 is 10×10, or 100. The exponent is 2 and it means that the 10 is multiplied by itself. The number 10^3 is $10 \times 10 \times 10$, or 1000.

The decimal part of scientific notation is a number between 1 and 10. The exponent on the 10 may be any whole number. For example, the number 123 is equivalent to 1.23×100. This is the same thing as saying that $123 = 1.23 \times 10^2$. The number 123.4 is equivalent to 1.234×10^2.

A simple way to keep track of the correct exponent is to move the decimal and count the number of places it is moved. For example, starting with the number 432.1, the decimal is to the right of the 2. We can move the decimal two places to the left to give us a decimal part that is between 1 and 10. This means that the exponent on the 10 will be a 2 and that $432.1 = 4.321 \times 10^2$.

Convert the following decimal numbers to numbers in scientific notation.

_____ **1.** 3785

_____ **2.** 28,312

_____ **3.** 264.2

_____ **4.** 65.10

_____ **5.** 3,526,000

Convert the following numbers in scientific notation to decimal numbers.

_____ **6.** 1.54×10^2

_____ **7.** 2.8312×10^4

_____ **8.** 7.34×10^4

_____ **9.** 5.643×10^6

_____ **10.** 8.635×10^2

A negative exponent represents the reciprocal of the number. For example, 10^{-2} is $1/(10 \times 10)$, or 0.01. The number 10^{-3} is $1/(10 \times 10 \times 10)$, or 0.001. This method is used to simplify the presentation of very small numbers. For example, 0.0123 is equivalent to 1.23×10^{-2}. The number 0.01234 is equivalent to 1.234×10^{-2}.

For very small numbers, we can also use the method of moving the decimal and counting the number of places it is moved. In this case, we move the decimal to the right so we end up with a negative exponent. For example, starting with the number 0.00001234, the decimal needs to be moved 5 places to the right to get the decimal part between 1 and 10. This means that $0.00001234 = 1.234 \times 10^{-5}$.

Convert the following decimal numbers to numbers in scientific notation.

_____ **11.** 0.01667

_____ **12.** 0.0002642

_____ **13.** 0.000060

_____ **14.** 0.00000000923

_____ **15.** 0.0100003

Convert the following numbers in scientific notation to decimal numbers.

_____ **16.** 2.271×10^{-2}

_____ **17.** 3.532×10^{-5}

_____ **18.** 7.9×10^{-3}

_____ **19.** 3.953×10^{-10}

_____ **20.** 5.000002×10^{-5}

Numbers in scientific notation can be added together or subtracted from each other as long as the exponents are the same. If the exponents are not the same, the numbers need to be converted to decimals to add or subtract them.

Add or subtract the following numbers as indicated.

_____ **21.** $1.54 \times 10^{2} + 2.63 \times 10^{2}$

_____ **22.** $6.83 \times 10^{-4} + 1.97 \times 10^{-4}$

_____ **23.** $5.932 \times 10^{2} + 6.48 \times 10^{2}$

_____ **24.** $5.34 \times 10^{2} - 3.41 \times 10^{2}$

_____ **25.** $1.65 \times 10^{9} - 6.31 \times 10^{2}$

Numbers in scientific notation can be multiplied or divided. For multiplication, the decimal parts of the number are multiplied together and the exponents are added together. For division, the decimal parts of the number are divided and the exponents are subtracted.

Multiply or divide the following numbers as indicated.

_____ **26.** $1.54 \times 10^{2} \times 2.63 \times 10^{2}$

_____ **27.** $2.83 \times 10^{-2} \times 1.97 \times 10^{-4}$

_____ **28.** $7.93 \times 10^{4} \times 6.02 \times 10^{-2}$

_____ **29.** $6.34 \times 10^{-3} \div 1.16 \times 10^{2}$

_____ **30.** $2.5 \times 10^{2} \div 4.0 \times 10^{-4}$

Name _____ Date _____

Activity 1-4—Interpolation

Many areas of instrumentation involve obtaining information from tables of data. When using tables, it is common that the known value to be used in the table (called the entry value) falls between the printed values. An interpolation is necessary to find the correct value between the nearest higher and lower values. The interpolation process is based on the assumption that there is a linear relationship between the higher and lower values.

An interpolation is done by subtracting the lower entry table value from the actual entry table value. This is divided by the difference between the higher and lower entry table values. The result is a fractional value that represents the location of the entry value relative to the upper and lower values. The desired value is obtained by multiplying the difference between the higher and lower desired table values by the entry fractional value, with the result added to the lower desired table value.

The procedure is used when the entry and desired values increase and decrease with each other. If the desired values increase or decrease in the opposite direction from the entry value, the result obtained in the last step should be subtracted from the higher value instead of being added to the lowest value as above. For example, a portion of a saturated steam table is as follows:

Pressure psia	Temp °F	Sp Vol ft³/lb
150.0	358.43	3.0139
120.0	341.27	3.7275

In this example, the steam pressure is 125.0 psig (139.7 psia). What is the equivalent saturation temperature and specific volume? The fractional value is 0.6567 [(139.7 − 120.0) ÷ (150.0 − 120.0) = 0.6567].

From the temperature column, the equivalent temperatures to the pressures of 120.0 psia and 150.0 psia are 341.27°F and 358.43°F. The difference between these numbers is 17.16°F. The fractional portion of the temperature difference is 11.27°F. This is calculated by multiplying this difference by the fractional value (17.16°F × 0.6567 = 11.27°F). Add this fractional value to the lower value to get the final equivalent temperature of 352.54°F (11.27°F + 341.27°F = 352.54°F).

The next step is to determine the equivalent value of the specific volume. The fractional value obtained in the first step is still used because we are still using the same pressure values from the table. The difference between the equivalent specific volumes for 120.0 psia and 150 psia is 0.7136 ft³/lb (3.7275 ft³/lb − 3.0139 ft³/lb = 0.7136 ft³/lb). This is multiplied by the fractional value to get the fractional portion of 0.4686 (0.7136 ft³/lb × 0.6567 ft³/lb = 0.4686 ft³/lb). This has to be subtracted from the greater value since the specific volume decreases as the pressure increases. This gives a specific volume of 3.2589 ft³/lb (3.7275 ft³/lb − 0.4686 ft³/lb = 3.2589 ft³/lb).

A pressure of 125 psig (139.7 psia) has an equivalent saturation temperature of 352.54°F and specific volume of 3.2589 ft³/lb.

A portion of a saturated steam table is shown below. Use the data to answer the following questions.

Pressure psia	Temp °F	Sp Vol ft³/lb
195.729	380.0	2.3353
153.010	360.0	2.9573

_____ **1.** The saturated steam temperature is 365.5°F. What is the equivalent pressure in psia?

_____ **2.** The saturated steam temperature is 365.5°F. What is the equivalent specific volume in ft³/lb?

SECTION
1 INTRODUCTION TO INSTRUMENTATION

chapter
2

Fundamentals of Process Control

REVIEW
QUESTIONS

Name _____   **Date** _____

T   F   **1.** Process control is a system that combines measuring and controlling instruments into an arrangement capable of automatic remote action.

_____   **2.** A(n) ___ variable is the dependent variable that is to be controlled in a control system.

_____   **3.** A(n) ___ variable is the independent variable in a process control system that is used to adjust the dependent variable.

_____   **4.** A ___ is the desired value at which the process should be controlled.
A. control variable
B. setpoint
C. process variable
D. manipulated variable

_____   **5.** A(n) ___ element is the sensing device that detects the condition of the process variable.

_____   **6.** A final element is a device that receives a control signal and regulates the amount of ___ or energy in a process.
A. material
B. delay time
C. temperature
D. heat

_____   **7.** ___ characteristics of an element describe the operation of the element at steady-state conditions when the process is not changing.
A. Static
B. Dynamic
C. Varying
D. Hysteresis

_____   **8.** ___ is the difference between the highest and lowest numbers in a measurement range.

_____   **9.** ___ is the degree to which an element provides the same result with successive occurrences of the same condition.
A. Sensitivity
B. Precision
C. Accuracy
D. Repeatability

T   F   **10.** Sensitivity is a gradual change in an element over time when the process conditions are constant.

_____ **11.** ___ is the time it takes an element to respond to a change in the value of the measured variable or to produce a change in the output signal due to a change in the input signal.

_____ **12.** ___ is a property of physical systems that do not react immediately to the forces applied to them or do not return completely to their original state.
    A. Hysteresis
    B. Fidelity
    C. Linearity
    D. Stability

_____ **13.** A(n) ___ is a control system that provides feedback to the controller on the state of the process variable.
    A. dynamic error
    B. closed loop
    C. open loop
    D. nonlinear response

_____ **14.** ___ control is the simplest and most common control strategy.
    A. Feedforward
    B. Proportional
    C. ON/OFF
    D. Open loop

_____ **15.** Proportional control is a control strategy that uses the ___ between the setpoint and the process variable.
    A. difference
    B. average
    C. maximum value
    D. minimum value

Name _____ **Date** _____

Activity 2-1—Dynamic Characteristics

Increasing			Decreasing	
% signal	**Pressure**		**% signal**	**Pressure**
0	0		100	50
10	1		90	49
20	2		80	48
30	8		70	42
40	14		60	36
50	20		50	30
60	26		40	24
70	32		30	18
80	38		20	12
90	44		10	6
100	50		0	0

Data is provided above on the changes in pressure as a control signal is changed.

1. Graph the data for both increasing and decreasing signals. Graph the % signal on the x-axis and the pressure on the y-axis.

2. What dynamic characteristic is illustrated?

   _____

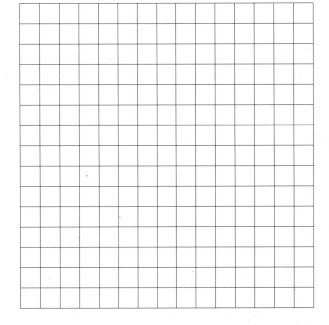

% signal	Temperature
0	0
10	20
20	42
30	58
40	70
50	75
60	80
70	100
80	110
90	125
100	150

Data is provided above on the changes in temperature as a control signal is changed.

3. Graph the data. Graph the % signal on the x-axis and the pressure on the y-axis.

4. What dynamic characteristic is illustrated?

_____

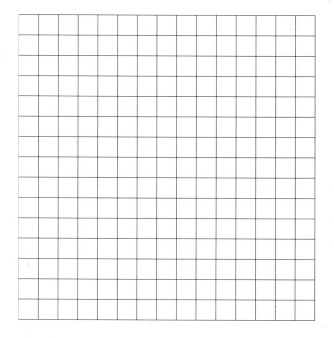

Name _____ **Date** _____

_____ 1. A ___ diagram is a schematic diagram of the relationship between instruments, controllers, piping, and system equipment.
   A. flow
   B. piping and instrumentation
   C. wiring
   D. ladder

_____ 2. A ___ is a circular symbol used to identify the purpose of an instrument or device.
   A. coder
   B. P&ID
   C. controller
   D. balloon

_____ 3. The letters TIC on a controller symbol on a P&ID stand for ___.
   A. temperature indicating controller
   B. totalized indicated change
   C. time identified current
   D. temperature increase current

_____ 4. A P&ID can show only part of the control strategy implemented in a ___ or a programmable logic controller.
   A. flow diagram
   B. relay wiring diagram
   C. ladder logic
   D. distributed control system

_____ 5. ___ has developed a comprehensive standard for instrumentation symbols.
   A. NEMA
   B. ISA
   C. OSHA
   D. IEEE

_____ 6. A P&ID uses the letter ___ for electrical current in an instrument tag balloon.
   A. P
   B. V
   C. E
   D. I

13

_____ **7.** A P&ID uses the letter ___ as a modifier representing an alarm in an instrument tag balloon.
     A. A
     B. D
     C. M
     D. X

_____ **8.** A P&ID uses the letter ___ for water flow in an instrument tag balloon.
     A. F
     B. Q
     C. W
     D. X

_____ **9.** The letters TSH on a P&ID represent "___."
     A. temperature switch high
     B. time safety helper
     C. trouble switch hourly
     D. temperature sensor half

_____ **10.** The letters TV on a P&ID represent "___."
     A. timed variation
     B. temperature valve
     C. thermocouple viscosity
     D. timed volume

Piping and Instrumentation Diagrams

Name _____   Date _____

Activity 3-1—Symbol Identification

Match the symbols to the description on the next page.

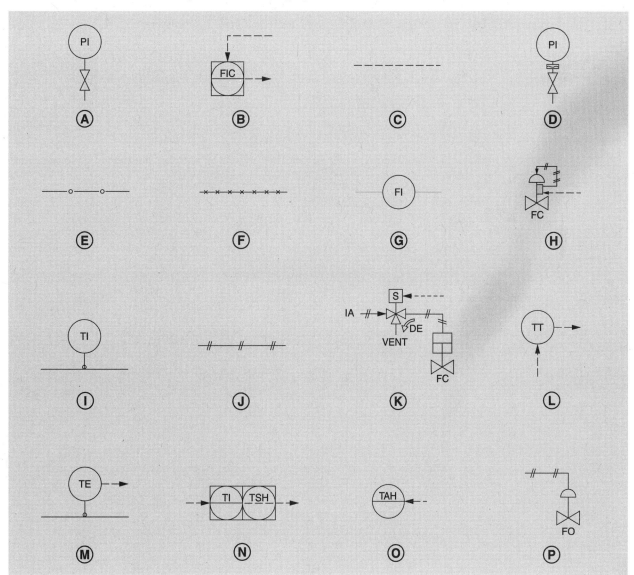

_____ 1. Fail closed piston-actuated ON/OFF valve actuated with a three-way solenoid valve

_____ 2. Fail open spring and diaphragm pneumatic control valve

_____ 3. Fail closed spring and diaphragm control valve with an electropneumatic positioner

_____ 4. Analog electrical transmission signal

_____ 5. Digital numerical value transmission signal

_____ 6. Pneumatic analog transmission signal

_____ 7. Direct-connected pressure gauge

_____ 8. Pressure gauge with a chemical seal

_____ 9. Local temperature indicator with a thermowell

_____ 10. Microprocessor panel-mounted flow indicating controller

_____ 11. In-line flow indicator

_____ 12. Electronic temperature transmitter with electrical measurement signal

_____ 13. High-temperature alarm annunciator

_____ 14. Electrical temperature element with a thermowell

_____ 15. Panel-mounted digital temperature indicator with an internal high-temperature alarm switch

_____ 16. Capillary tube filled with fluid

SECTION
2 TEMPERATURE MEASUREMENT

chapter
4

Temperature, Heat, and Energy

REVIEW
QUESTIONS

Name _____ **Date** _____

T   F   **1.** Conduction is a heat transfer method used to remove heat from sensitive electronic parts.

T   F   **2.** A reference temperature is a known fixed point needed to define a temperature scale.

T   F   **3.** A calorie is the amount of energy necessary to change the temperature of 1 lb of water by 1°F from 59°F to 60°F.

_____   **4.** ___ is the degree or intensity of heat measured on a definite scale.

_____   **5.** ___ is the unaided movement of a gas or liquid caused by a pressure difference due to a difference in density within the gas or liquid.
A. Conduction
B. Induction
C. Forced convection
D. Natural convection

_____   **6.** ___ is the lowest temperature possible where there is no molecular movement and the energy is at a minimum.
A. Absolute zero
B. Positive zero
C. Negative 100
D. Absolute 100

_____   **7.** Four common ___ are the Fahrenheit, Rankine, Celsius, and Kelvin.
A. units of length
B. constants
C. thermocouples
D. temperature scales

_____   **8.** The ___ of a material is the amount of energy needed to change the temperature of the material by a certain amount.
A. conduction
B. heat capacity
C. radiation
D. temperature

T   F   **9.** A heat sink is a heat conductor that adds heat to electronic parts by pressure difference.

T   F   **10.** Heat transfer is the movement of thermal energy from one place to another.

_____ **11.** ___ are poor conductors of heat.
    A. Insulators
    B. Transmitters
    C. Metals
    D. all of the above

T   F **12.** Thermal equilibrium is the state in which objects are at the same temperature and there is no heat transfer between them.

_____ **13.** ___ is heat transfer that occurs when molecules in a material are heated and the heat is passed from molecule to molecule through the material.

_____ **14.** The ___ of a liquid is the ratio of the heat capacity of that liquid to the heat capacity of water at the same temperature.
    A. specific heat
    B. temperature
    C. volume
    D. meniscus

_____ **15.** ___ is the movement of a gas or liquid due to a pressure difference caused by the mechanical action of a fan or pump.
    A. Natural convection
    B. Forced convection
    C. Induction
    D. Conduction

T   F **16.** Metals are generally good conductors of heat.

_____ **17.** ___ is heat transfer by electromagnetic waves emitted by a higher-temperature object and absorbed by a lower-temperature object.

_____ **18.** A(n) ___ is an instrument that is used to indicate temperature.

T   F **19.** Convective energy waves move through air or space while producing heat.

_____ **20.** A ___ is the amount of energy necessary to change the temperature of 1 g of water by 1°C from 14.5°C to 15.5°C.
    A. British thermal unit (Btu)
    B. calorie (cal)
    C. Kelvin
    D. none of the above

_____ **21.** Heat is transferred by ___ from a flame into a boiler to create steam.
    A. conduction
    B. convection
    C. radiation
    D. all of the above

T   F **22.** Triple point is the condition where all three phases of a substance—gas, liquid, and solid—can coexist in equilibrium.

_____ **23.** A(n) ___ is the time required for a thermometer to change by 63.2% of its total change in temperature.

Temperature, Heat, and Energy

Name _____   **Date** _____

Activity 4-1—Temperature Conversions

For each of the following problems, convert the given temperature to the equivalent value in each of the other temperature scales. Round answers to the nearest degree.

_____   **1.** 104°F = ___°C

_____   **2.** 37°C = ___°F

_____   **3.** 185°F = ___°C

_____   **4.** 232°C = ___ K

_____   **5.** 68°F = ___°R

_____   **6.** 373 K = ___°F

_____   **7.** 77 K = ___°C

_____   **8.** –109°F = ___ K

_____   **9.** 559°R = ___ K

_____   **10.** 180°C = ___°F

For each of the following questions, calculate the value as requested.

_____   **11.** What is the temperature at which the Fahrenheit and Celsius temperature scales read the same value?

_____   **12.** What is the temperature of an object where it is half as hot as 80°F?

Name _____ Date _____

Activity 4-2—Heat Capacity

Heat capacity is the amount of heat energy in Btu required to change the temperature of 1 lb of material by 1°F. There is a linear relationship between the heat transferred and the temperature as long as the material does not change phase. The heat added to change the temperature of the material is called sensible heat. Liquids can freeze when heat is lost and vaporize when heat is applied. At these points of phase change, considerable amounts of energy must be either removed or added for the material to completely pass through the phase change. It takes 143.5 Btu/lb for water to pass through the freezing point. This is called heat of fusion. It takes 970.3 Btu/lb for water to be vaporized at atmospheric pressure. This is called heat of vaporization. The heat of vaporization decreases with increases in pressure.

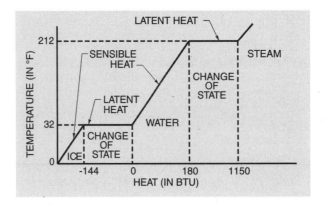

For each of the following questions, calculate the heat energy.

_____ 1. What is the amount of heat energy in Btu that must be added to a gallon of water at 72°F to raise the temperature to 212°F?

_____ 2. What is the amount of heat energy in Btu that must be added to a gallon of water at 212°F to convert it to steam at the same temperature?

_____ 3. What is the amount of heat energy in Btu that must be removed from a gallon of water at 32°F to convert it to ice at the same temperature?

_____ 4. What is the amount of heat energy in Btu that must be removed from a gallon of water at 72°F to lower the temperature to 32°F?

Name _____ **Date** _____

Activity 4-3—Heat Transfer

A chilled water heat exchanger uses brine with a specific gravity of 1.2 to cool water. The brine inlet temperature is 0°F and its outlet temperature is 35°F. The heat capacity of the brine is 0.79 Btu/lb°F and the brine flow rate is 25 gpm. The water outlet temperature is controlled at 35°F and the water inlet temperature is 55°F. The heat capacity of water is 1.0 Btu/lb°F.

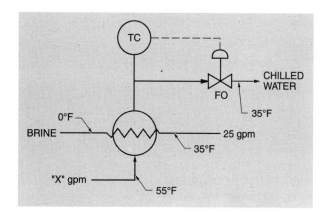

For each of the following questions, calculate the quantities as requested.

_____ **1.** What is the mass flow rate of the brine in lb/min?

_____ **2.** What is the amount of heat transferred to the brine in Btu/min?

_____ **3.** What is the mass flow rate of the water in lb/min?

_____ **4.** What is the flow of the water in gpm?

Temperature, Heat, and Energy

ACTIVITIES

Name _____ Date _____

Activity 4-4—Steam Tables

Many industrial processes are based on the use of water and steam. Saturated steam is steam at its condensing temperature. Saturated steam tables list the properties of saturated steam related to pressure or temperature. Specific volume is the reciprocal of density. The tables list the specific volume in ft³/lb for water at its boiling point and for steam at its saturation point. Enthalpy is the amount of heat energy stored in the water or steam. The tables list the enthalpy in Btu/lb for water, steam, and the change in heat energy to go from one to the other, labeled "Evap." The third set of data is entropy, which is rarely used in industry.

Dry Saturated Steam Pressure									
		Specified Volume‡		Enthalpy§			Entropy		
Absolute Pressure* p	Temperature† t	Saturated Liquid V_f	Saturated Vapor V_g	Saturated Liquid h_f	Evap. h_{fg}	Saturated Vapor h_g	Saturated Liquid S_f	Evap. S_{fg}	Saturated Vapor S_g
0.122	40.0	0.01602	2445.8	8.03	1071.0	1079.0	0.0162	2.1432	2.1594
0.256	60.0	0.01603	1207.6	28.06	1059.7	1087.7	0.0555	2.0391	2.0946
0.507	80.0	0.01607	633.3	48.04	1048.4	1096.4	0.0932	1.9426	2.0359
0.949	100.00	0.01613	350.4	68.00	1037.1	1105.1	0.1295	1.8530	1.9825
1	101.74	0.01614	333.6	69.70	1036.3	1106.0	0.1326	1.8456	1.9782
140	353.02	0.01802	3.220	324.82	868.2	1193.0	0.5069	1.0682	1.5751
150	358.42	0.01809	3.015	330.51	863.6	1194.1	0.5138	1.0556	1.5694
160	363.53	0.01815	2.834	335.93	859.2	1195.1	0.5204	1.0436	1.5640
170	368.41	0.01822	2.675	341.09	854.9	1196.0	0.5266	1.0324	1.5590

* in psia
† in °F
‡ in cu ft/lb
§ in btu/lb

A boiler is operated at 150 psia (135.3 psig) and produces 25,000 lb/hr of steam. Calculate the amount of heat energy in Btu/hr that is needed to produce the required steam when the feedwater temperature is 60°F.

_____ 1. What is the amount of heat energy in Btu/lb contained in the water at 60°F?

_____ 2. What is the amount of heat energy in Btu/lb contained in the steam at 150 psia?

_____ 3. What is the amount of heat energy in Btu/lb required to convert the water to steam?

_____ 4. What is the amount of heat energy in Btu/hr required to produce 25,000 lb/hr of steam?

Name _____ **Date** _____

T   F   **1.** A bimetallic thermometer is another name for a pressure-spring thermometer.

_____   **2.** Common thermal expansion instruments are ___ thermometers.
     A. liquid-in-glass
     B. bimetallic
     C. pressure-spring
     D. all of the above

_____   **3.** A(n) ___ thermometer is a thermal expansion thermometer consisting of a sealed, narrow-bore glass tube with a bulb at the bottom and filled with a liquid.

_____   **4.** The temperature response time when measuring a liquid with a thermal expansion thermometer is ___ when measuring a gas.
     A. slower than
     B. faster than
     C. the same as
     D. undetermined

T   F   **5.** For accurate temperature measurement, a thermometer should always be immersed in a material just far enough to completely cover the bulb.

T   F   **6.** The liquid used in liquid-in-glass thermometers is usually mercury, alcohol, or an organic liquid with a red dye added to improve visibility.

_____   **7.** With a(n) ___ thermometer, graduations are engraved on metal plates rather than marked or scaled on glass tubes.
     A. infrared
     B. industrial
     C. bimetallic
     D. none of the above

_____   **8.** Alloys that have closely controlled coefficients of thermal expansion make a ___ thermometer a very dependable temperature-measuring device.
     A. liquid-in-glass
     B. bimetallic
     C. pressure-spring
     D. none of the above

T   F   **9.** A bimetallic temperature element can be used as a temperature switch.

_____ **10.** A(n) ___ is a secondary chamber fitted into the metal tube chamber.
    A. element
    B. coefficient of thermal expansion
    C. separable socket
    D. none of the above

_____ **11.** A ___ thermometer is a pressure-spring thermometer that measures the increase in pressure of a confined (kept at constant volume) gas due to a temperature increase.
    A. gas-filled pressure-spring
    B. liquid-in-glass
    C. bimetallic
    D. vapor-pressure pressure-spring

_____ **12.** A ___ thermometer is a pressure-spring thermometer that uses the change in vapor pressure due to the temperature change of an organic liquid to determine the temperature.
    A. bimetallic
    B. gas-filled pressure-spring
    C. vapor-pressure pressure-spring
    D. liquid-in-glass

_____ **13.** A bimetallic thermometer is a thermal expansion thermometer that uses ___ to measure temperature.
    A. a strip consisting of two alloys
    B. infrared radiation
    C. the coefficient of friction
    D. all of the above

_____ **14.** A(n) ___ consists of a bimetallic strip that is usually wound into a spiral, helix, or coil to allow the instrument to be placed in a smaller space than the straight element requires.

_____ **15.** The difference in height between the ___ and the pressure spring of a thermometer can introduce error.

T   F   **16.** In applications that use a bimetallic element to make a measurement over a small temperature range, alloys with widely differing rates of thermal expansion are used.

_____ **17.** Bimetallic thermometers can be used to measure temperatures from ___°F to ___°F.
    A. –300; 800
    B. –300; 100
    C. –112; 760
    D. –100; 500

T   F   **18.** A large number of reference marks enables a liquid-in-glass thermometer to be read more accurately than one with a small number of reference marks.

_____ **19.** A(n) ___ is a thermal expansion thermometer consisting of a filled, hollow spring attached to a capillary tube and bulb where the fluid in the bulb expands or contracts with temperature changes.

T   F   **20.** A vapor-pressure pressure-spring thermometer is a pressure-spring thermometer that is filled with mercury under pressure.

Name _____ **Date** _____

Activity 5-1—Instrument Connections

Four piping arrangements are shown below. The two on the left require instrument connections into ¾″ threaded or half couplings 1″ long welded to the pipe. The two on the right require instrument connections into 1″-150 lb flanged connections 3″ long. The flanged connections allow room for insulation around the pipe. In each pair, the top one shows an expanded section in the elbow when the main pipe is smaller than 3″ to allow room for instruments. For each problem below, all dial thermometers have a 5″ dial, union connections, bimetallic element, ¼″ diameter stem, and a ½″ NPT-M connection.

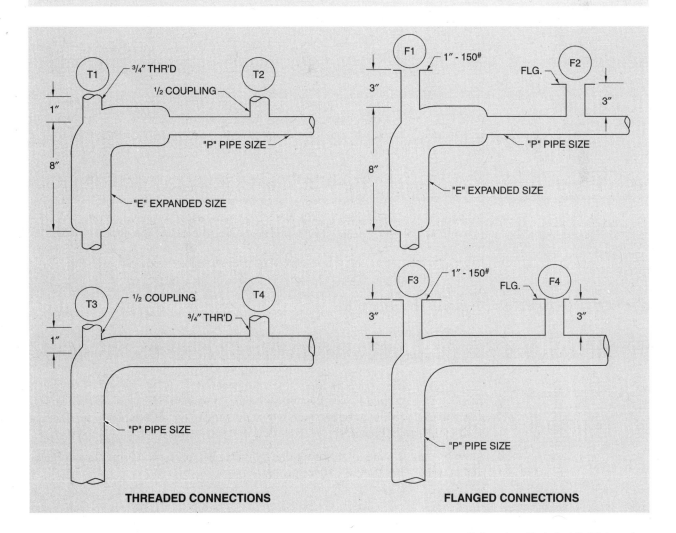

THREADED CONNECTIONS   FLANGED CONNECTIONS

Given the process connection size and type, pipe size, and whether the pipe is to be insulated or not, select the best location for the thermometer, the expanded pipe size "E," insertion length "U," lagging extension length "T," and dial thermometer stem length "S."

¾" NPT connection, 2" pipe, no insulation

_____ **1.** The best location is ___.

_____ **2.** The expanded pipe size is ___.

_____ **3.** The insertion length is ___.

_____ **4.** The lagging extension length is ___.

_____ **5.** The stem length is ___.

¾" NPT connection, 3" pipe, 2" insulation

_____ **6.** The best location is ___.

_____ **7.** The insertion length is ___.

_____ **8.** The lagging extension length is ___.

_____ **9.** The stem length is ___.

1"-150 lb flange connection, 2" pipe, 2" insulation

_____ **10.** The best location is ___.

_____ **11.** The expanded pipe size is ___.

_____ **12.** The insertion length is ___.

_____ **13.** The lagging extension length is ___.

_____ **14.** The stem length is ___.

1"-150 lb flange connection, 4" pipe, no insulation

_____ **15.** The best location is ___.

_____ **16.** The insertion length is ___.

_____ **17.** The lagging extension length is ___.

_____ **18.** The stem length is ___.

SECTION
2 TEMPERATURE MEASUREMENT

chapter
6

Electrical Thermometers

REVIEW
QUESTIONS

Name _____   **Date** _____

_____   1. ___ is the process by which thermocouples gradually change their voltage-temperature curve due to extended time in extreme environments.
   A. The Seebeck effect
   B. Thermocouple aging
   C. Resistance bridging
   D. The Peltier effect

_____   2. The ___ is the joined end of the thermocouple that is exposed to the process where the temperature measurement is desired.
   A. hot junction
   B. thermometer
   C. cold junction
   D. isothermal block

T   F   3. A Seebeck voltage cannot be measured directly.

T   F   4. A thermocouple junction is a point where the two dissimilar wires are joined.

_____   5. The ___ is a thermoelectric effect where heating or cooling occurs at the junctions of two dissimilar conductive materials when a current flows through the junctions.

T   F   6. Shunt impedance is a circuit that is used to precisely measure an unknown resistance.

_____   7. ___ is the process of using automatic compensation with a thermocouple to calculate temperatures when the reference junction is not at the ice point.

_____   8. A(n) ___ is an electrical thermometer consisting of several thermocouples connected in series to provide a higher voltage output.
   A. cold junction compensation
   B. ground loop
   C. thermopile
   D. RTD

T   F   9. A break point is the temperature at which the resistance of an RTD begins to increase rapidly.

_____ **10.** A(n) ___ is an electrical thermometer consisting of a set of parallel-connected thermocouples that is commonly used to measure an average temperature of an object or area.
     A. thermopile
     B. averaging thermocouple
     C. hot junction
     D. cold junction

T   F   **11.** The law of intermediate metals states that the use of a third metal in a thermocouple circuit does not affect the voltage, as long as the temperature of the three metals at the point of junction is the same.

T   F   **12.** The Seebeck effect is a thermoelectric effect where continuous current is generated in a circuit where the junctions of two dissimilar conductive materials are kept at different temperatures.

_____ **13.** A(n) ___ is a thermometer consisting of a high-precision resistor with resistance that varies with temperature, a voltage or current source, and a measuring circuit.

T   F   **14.** A null-current thermocouple consists of a circuit and a voltage generator that can be adjusted to exactly balance the voltage output of the thermocouple.

_____ **15.** The ___ is the end of a thermocouple that is kept at a constant temperature in order to provide a reference point.
     A. hot junction
     B. thermopile
     C. cold junction
     D. none of the above

T   F   **16.** A difference thermocouple is a pair of thermocouples connected together to measure a temperature difference between two objects.

T   F   **17.** A thermocouple is a temperature-sensitive resistor consisting of solid-state semiconductors made from sintered metal oxides and lead wires, hermetically sealed in glass.

_____ **18.** A(n) ___ is an electrical thermometer consisting of two dissimilar metal wires joined at one end and a voltmeter to measure the voltage at the other end of the two wires.

Name _____   Date _____

Activity 6-1—Thermocouple Voltages

Modern thermocouple calibration instruments automatically adjust to maintain a 32°F (0°C) ice point reference for both reading and simulating thermocouples. Although this is automatic, it is important to understand the basis for these automatic operations. The values shown are in millivolts referenced to the 32°F (0°C) mV value. The voltage produced at the ambient temperature must be added to the voltage produced at the elevated temperature.

Type J Thermoelectric Voltage in mV

°C	0	−1	−2	−3	−4	−5	−6	−7	−8	−9	−10
−40	−1.961	−2.008	−2.055	−2.103	−2.150	−2.197	−2.244	−2.291	−2.338	−2.385	−2.431
−30	−1.482	−1.530	−1.578	−1.626	−1.674	−1.722	−1.770	−1.818	−1.865	−1.913	−1.961
−20	−0.995	−1.044	−1.093	−1.142	−1.190	−1.239	−1.288	−1.336	−1.385	−1.433	−1.482
−10	−0.501	−0.550	−0.600	−0.650	−0.699	−0.749	−0.798	−0.847	−0.896	−0.946	−0.995
0	0.000	−0.050	−0.101	−0.151	−0.201	−0.251	−0.301	−0.351	−0.401	−0.451	−0.501

°C	0	1	2	3	4	5	6	7	8	9	10
0	0.000	0.050	0.101	0.151	0.202	0.253	0.303	0.354	0.405	0.456	0.507
10	0.507	0.558	0.609	0.660	0.711	0.762	0.814	0.865	0.916	0.968	1.019
20	1.019	1.071	1.122	1.174	1.226	1.277	1.329	1.381	1.433	1.485	1.537
30	1.537	1.589	1.641	1.693	1.745	1.797	1.849	1.902	1.954	2.006	2.059
40	2.059	2.111	2.164	2.216	2.269	2.322	2.374	2.427	2.480	2.532	2.585
50	2.585	2.638	2.691	2.744	2.797	2.850	2.903	2.956	3.009	3.062	3.116
60	3.116	3.169	3.222	3.275	3.329	3.382	3.436	3.489	3.543	3.596	3.650
70	3.650	3.703	3.757	3.810	3.864	3.918	3.971	4.025	4.079	4.133	4.187
80	4.187	4.240	4.294	4.348	4.402	4.456	4.510	4.564	4.618	4.672	4.726
90	4.726	4.781	4.835	4.889	4.943	4.997	5.052	5.106	5.160	5.215	5.269

The following mV readings were made with a digital voltmeter with a Type J thermocouple and no ice point compensation. The ambient temperature is provided. Determine the actual temperature at the hot junction.

_____ 1. Given an ambient temperature of 0°C and a voltage reading of +1.277 mV, the hot junction temperature is ___.

_____ 2. Given an ambient temperature of 15°C and a voltage reading of +4.181 mV, the hot junction temperature is ___.

_____ 3. Given an ambient temperature of 20°C and a voltage reading of −0.154 mV, the hot junction temperature is ___.

_____ 4. Given an ambient temperature of 25°C and a voltage reading of −2.173 mV, the hot junction temperature is ___.

_____ 5. Given an ambient temperature of −10°C and a voltage reading of +5.336 mV, the hot junction temperature is ___.

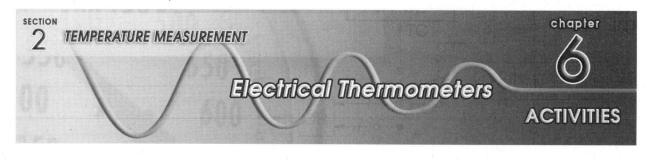

SECTION
2 TEMPERATURE MEASUREMENT

Electrical Thermometers

chapter
6

ACTIVITIES

Name _____ Date _____

Activity 6-2—Thermocouple Circuits

1. Draw a simple null current circuit for measuring thermocouple temperatures. Do not use an ice point reference in the circuit.

2. Draw a thermocouple circuit that can be used to average three temperatures.

3. Draw a thermocouple circuit that can be used to measure a temperature difference between two objects.

Name _____   **Date** _____

Activity 6-3—RTD Tables

RTDs are precision resistance elements that are used to measure temperature more accurately than thermocouples. There are a number of different RTD temperature/resistance tables. Some are based on 100 Ω at 0°C. Nickel RTD elements have different curves than platinum elements. The two most common 100 Ω platinum RTD types are the "American curve," alpha = 0.00392, and the "European curve," alpha = 0.00385. The European RTD curve is commonly used in the United States.

Use the RTD table in the Appendix to determine the temperature equivalents for the following resistances.

_____ **1.** Given a resistance of 84.27 Ω, the temperature is ___°F.

_____ **2.** Given a resistance of 114.04 Ω, the temperature is ___°F.

_____ **3.** Given a resistance of 138.51 Ω, the temperature is ___°F.

_____ **4.** Given a resistance of 208.48 Ω, the temperature is ___°F.

_____ **5.** Given a resistance of 104.66 Ω, the temperature is ___°F.

A portable RTD calibrator is used as an input to a transmitter to confirm the calibration ranges. Use the RTD table to determine the resistance values needed to represent the following temperature values for an RTD used to measure water temperature.

_____ **6.** Given a temperature of 32°F, the resistance is ___ Ω.

_____ **7.** Given a temperature of 70°F, the resistance is ___ Ω.

_____ **8.** Given a temperature of 110°F, the resistance is ___ Ω.

_____ **9.** Given a temperature of 150°F, the resistance is ___ Ω.

_____ **10.** Given a temperature of 190°F, the resistance is ___ Ω.

_____ **11.** Given a temperature of 212°F, the resistance is ___ Ω.

A portable RTD calibrator is used as an input to a transmitter to confirm the calibration ranges. Use the RTD table to determine the resistance values needed to represent the following temperature values for an RTD used to measure the temperature of a chemical reactor.

_____ 12. Given a temperature of 0°F, the resistance is ___ Ω.

_____ 13. Given a temperature of 100°F, the resistance is ___ Ω.

_____ 14. Given a temperature of 200°F, the resistance is ___ Ω.

_____ 15. Given a temperature of 300°F, the resistance is ___ Ω.

_____ 16. Given a temperature of 400°F, the resistance is ___ Ω.

_____ 17. Given a temperature of 500°F, the resistance is ___ Ω.

_____ 18. Given a temperature of 600°F, the resistance is ___ Ω.

SECTION
2 TEMPERATURE MEASUREMENT

chapter
7

Infrared Radiation Thermometers

REVIEW
QUESTIONS

Name _____   **Date** _____

T   F   **1.** A two-color IR thermometer is independent of the emissivity of the surface as long as the emissivity of a measured surface is the same for both wavelengths of the two IR detectors.

T   F   **2.** The word "color" in one-color or two-color IR thermometer is misleading because these thermometers measure infrared radiation instead of visible light.

_____   **3.** A(n) ___ thermometer measures the infrared radiation emitted by an object to determine the object's temperature.

T   F   **4.** A radiometric thermal imager uses a rotating mirror with a single, very fast response detector or a linear array of IR detectors to measure successive areas with a single stationary device.

_____   **5.** The ratio ___ is often used to describe optical resolution.
      A. distance:size
      B. distance:intensity
      C. dimension:spot
      D. spot:dimension

T   F   **6.** A line scanner IR thermometer is an imager where the temperature measurement at all positions in the image is known.

_____   **7.** ___ is the range of infrared wavelengths measured by an IR thermometer.

_____   **8.** The wavelength and amplitude of the radiation emitted by a body changes with the ___ of the body.
      A. temperature
      B. distance
      C. size
      D. shape

T   F   **9.** A whitebody reflects a portion of all wavelengths of radiation equally.

_____   **10.** A(n) ___ thermometer has two IR detectors that measure infrared radiation at two different wavelengths.

T   F   **11.** Bodies that are at thermal equilibrium must balance the energy entering that object with the energy leaving that object.

_____   **12.** A(n) ___ is a high-temperature thermometer that has an electrically heated, calibrated tungsten filament contained within a telescope tube.

_____ **13.** A bright, shiny object that reflects light generally has a ___ emissivity.
    A. high
    B. low
    C. negative
    D. none of the above

_____ **14.** Infrared radiation thermometers generally have ___ response times than thermal expansion thermometers.
    A. slower
    B. faster
    C. equal
    D. cannot be determined

_____ **15.** ___ is the ability of objects like glass to allow infrared radiation to pass through.

T   F   **16.** A background temperature thermometer measures infrared radiation using one IR detector.

T   F   **17.** A disappearing filament pyrometer can only be used with surfaces that have a visible color change above about 1000°F.

_____ **18.** ___ is the ability of a body to reflect thermal energy.

_____ **19.** ___ is an adjustment to the relationship between two IR detectors to compensate for a difference in infrared radiation detected.
    A. Sensitivity
    B. Frequency
    C. Slope
    D. Wavelength

_____ **20.** A ___ is an ideal body that completely absorbs all radiant energy of any wavelength falling on it and reflects none of this energy from the surface.
    A. redbody
    B. blackbody
    C. graybody
    D. whitebody

T   F   **21.** Radiant heat can be felt through a window because of the emissivity of the window.

_____ **22.** A(n) ___ is a temperature-measuring instrument that is used to measure temperatures beyond the range of a mercury thermometer.
    A. element
    B. blackbody
    C. graybody
    D. pyrometer

T   F   **23.** The infrared segment of the electromagnetic spectrum is not visible, but the wavelength of the radiation can be measured.

_____ **24.** A ___ is an infrared device that uses a two-dimensional array of IR detectors to generate a thermal image.
    A. line scanner IR thermometer
    B. radiometric thermal imager
    C. thermal imager
    D. pyrometer

Name _____ Date _____

Activity 7-1—Background Temperature Effects

The temperature of molten glass from a glass furnace is measured with a fixed IR detector and sent to a controller with a temperature transmitter. The molten glass stream is narrow as it exits the furnace. Due to the radiated heat from the furnace and molten glass, the IR detector must be mounted a considerable distance from the furnace. The detector's D:S ratio is such that the detector spot is larger than the target at that distance. The signal that the detector picks up is 75% from the molten glass and 25% from the background. The molten glass has an emissivity of 0.98. The background is 400°F with an emissivity of 0.9.

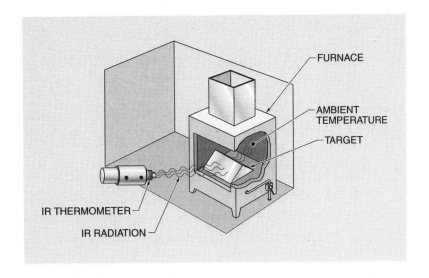

_____ 1. If the measured temperature is 1200°F, what is the actual molten glass temperature?

_____   **2.** List at least three things that can be done to reduce the measurement error.

SECTION
2 TEMPERATURE MEASUREMENT

chapter
8

Practical Temperature
Measurement and Calibration

REVIEW
QUESTIONS

Name _____   **Date** _____

_____ 1. ___ temperature-measuring instruments need to be calibrated.
   A. All
   B. No
   C. Only specific types of
   D. none of the above

_____ 2. Infrared radiation thermometers are calibrated with a ___ calibrator.
   A. thermal
   B. blackbody
   C. graybody
   D. all of the above

_____ 3. A(n) ___ is a temperature-controlled box where a thermometer can be inserted
   and the output compared to the known temperature.
   A. dry well calibrator
   B. water bath
   C. blackbody
   D. electronic calibrator

_____ 4. A ___ is a small tank containing a stirred liquid used to calibrate thermometers.
   A. blackbody
   B. dry well
   C. microbath
   D. none of the above

_____ 5. A(n) ___ is a device used to calibrate infrared thermometers.
   A. transmission calibrator
   B. emissivity calibrator
   C. blackbody calibrator
   D. dry well

_____ 6. A standard temperature transmitter can be calibrated by removing the input wires
   and replacing them with a(n) ___.
   A. blackbody calibrator
   B. electronic calibrator
   C. RTD
   D. thermocouple

_____ 7. The actual temperature of a dry well block or microbath is measured with a(n) ___ thermometer.
    A. reference
    B. infrared
    C. blackbody
    D. graybody

_____ 8. An electronic calibrator generates a(n) ___ that replicates the signal from an electrical thermometer.
    A. electrical signal
    B. 4 mA to 20 mA signal
    C. blackbody signal
    D. resistance

_____ 9. Many transmitters include a(n) ___ to eliminate an electrical path for ground currents and other electrical noise.
    A. earth ground
    B. dynamic calibrator
    C. optical isolator
    D. all of the above

_____ 10. Blackbody calibrators have a known ___ of nearly 1.0.
    A. transmissivity
    B. emissivity
    C. graybody
    D. whitebody

_____ 11. A(n) ___ resistance bridge has fixed resistances and the voltage across the bridge is proportional to the temperature of the variable resistance device.

_____ 12. A(n) ___ is a circuit that has more than one point connected to the earth ground, with a voltage potential difference between the two ground points high enough to produce a circulating current in the ground system.

_____ 13. Thermocouple break protection is a circuit where an electronic device sends ___ current across the thermocouple.
    A. no
    B. high-level
    C. low-level
    D. none of the above

T   F 14. A ground loop is an unintended circuit caused by a breakdown of thermocouple insulation.

_____ 15. A(n) ___ resistance bridge has variable resistances that are adjusted so there is equal current flow through the legs of the bridge and zero potential across the bridge.

Name _____   Date _____

Activity 8-1—Thermocouple Calibration

A millivolt generation instrument, without ice point compensation, can be used to provide thermocouple input signals for the calibration of temperature transmitters. The ambient temperatures are provided. Determine the mV output to obtain the stated calibration point temperature. Use the thermocouple voltage table in the Appendix for the Type J thermocouple and the thermocouple voltage table in the Appendix for the Type K thermocouple.

Given a Type J thermocouple and an ambient temperature of 20°C, answer the following questions.

_____ 1. The calibration voltage input signal at 0°C is ___ mV.

_____ 2. The calibration voltage input signal at 100°C is ___ mV.

_____ 3. The calibration voltage input signal at 200°C is ___ mV.

_____ 4. The calibration voltage input signal at 300°C is ___ mV.

_____ 5. The calibration voltage input signal at 400°C is ___ mV.

Given a Type K thermocouple and an ambient temperature of 26°C, answer the following questions.

_____ 6. The calibration voltage input signal at 0°C is ___ mV.

_____ 7. The calibration voltage input signal at 100°C is ___ mV.

_____ 8. The calibration voltage input signal at 200°C is ___ mV.

_____ 9. The calibration voltage input signal at 300°C is ___ mV.

_____ 10. The calibration voltage input signal at 400°C is ___ mV.

Given a Type K thermocouple and an ambient temperature of 35°C, answer the following questions.

_____ 11. The calibration voltage input signal at 0°C is _____ mV.

_____ 12. The calibration voltage input signal at 100°C is _____ mV.

_____ 13. The calibration voltage input signal at 200°C is _____ mV.

_____ 14. The calibration voltage input signal at 300°C is _____ mV.

_____ 15. The calibration voltage input signal at 400°C is _____ mV.

Name _____   Date _____

Activity 8-2—RTD Bridge Circuits

The resistance of an RTD is measured with a Wheatstone bridge circuit. To eliminate errors from wiring resistance inaccuracy, RTDs use matched lead wires in a 3-wire or 4-wire bridge circuit.

1. Draw a simple 2-wire Wheatstone bridge circuit.

2. Draw a simple 3-wire Wheatstone bridge circuit.

3. Draw a simple 4-wire Wheatstone bridge circuit.

4. Explain the advantage of using a 3-wire or 4-wire Wheatstone bridge circuit instead of a 2-wire circuit.

TEMPERATURE MEASUREMENT

Name _____   **Date** _____

Pyrometers

A pyrometer is an electrical temperature-measuring instrument powered only by a thermocouple. Assume that the meter is an ammeter with a 1 mA maximum range and a meter resistance of 1 Ω. The meter is kept at 70°F. The scale of the meter is 70°F to 250°F. It is intended to use a Type J 14-ga thermocouple and extension wire. The distance from the thermocouple to the meter is 35 ft and the Type J resistance is 0.09 Ω/double foot.

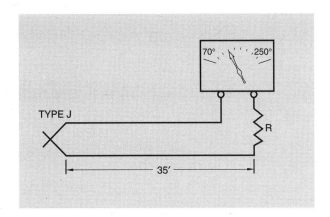

_____   **1.** How much resistance must be put into the circuit to make the meter read the correct temperature?

SECTION
3 *PRESSURE MEASUREMENT*

chapter
9

Pressure

REVIEW
QUESTIONS

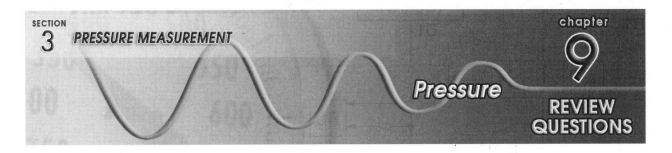

Name _____ **Date** _____

_____ **1.** ___ pressure is the pressure due to the head of a liquid column.
  A. Hydrostatic
  B. Absolute
  C. Gauge
  D. Differential

T F **2.** Force is the number of unit squares equal to the surface of an object.

_____ **3.** ___ pressure is due to the weight of the atmosphere above the point where it is measured.

T F **4.** Pressure is anything that changes or tends to change the state of rest or motion of a body.

T F **5.** Pressure is independent of the shape of a container and depends only on the properties of fluid and height.

_____ **6.** A pressure of 1 psi is equal to a hydrostatic head of ___ inches of water.
  A. 14.7
  B. 23.5
  C. 27.7
  D. 30.2

_____ **7.** ___ pressure is measured with a perfect vacuum as the zero point of the scale.
  A. Absolute
  B. Vacuum
  C. Differential
  D. Gauge

_____ **8.** A ___ is any material that tends to flow.
  A. pressure
  B. pascal
  C. pressure difference
  D. fluid

_____ **9.** The head of a column of liquid can be related to the actual pressure based on the height and density of the liquid by the formula ___.
  A. $P = \rho \times h$
  B. $P = h \div \rho$
  C. $P = \rho \div h$
  D. none of the above

T   F   **10.** Absolute zero pressure is a perfect vacuum.

_____ **11.** ___ is force divided by the area over which that force is applied.
                            A. Fluid
                            B. Pressure
                            C. Gauge
                            D. Pascal

T   F   **12.** Hydraulic pressure is the pressure of air or another gas that is compressed.

T   F   **13.** Head is the actual height of a column of liquid.

T   F   **14.** At mean sea level, the standard pressure of air is about 27.7 psi.

_____ **15.** ___ pressure is measured with atmospheric pressure as the zero point of the scale.
                            A. Gauge
                            B. Hydrostatic
                            C. Absolute
                            D. Hydraulic

T   F   **16.** A 1′ column of water has a pressure of 0.433 psi, which means that water has a hydrostatic pressure of 0.433 psi/ft of height.

T   F   **17.** Differential pressure is the difference in pressure between two measurement points in a process.

_____ **18.** ___ law states that pressure applied to a confined static fluid is transmitted with equal intensity throughout the fluid.
                            A. Boyle's
                            B. Newton's
                            C. Pascal's
                            D. Dalton's

_____ **19.** When a pressure of 200 psi is transmitted equally from one hydraulic cylinder to a load cylinder, the force on the load cylinder piston with an area of 5 sq in. is ___ lb.
                            A. 40
                            B. 200
                            C. 500
                            D. 1000

T   F   **20.** Pneumatic pressure is the pressure in a confined hydraulic liquid that has been subjected to the action of a pump.

_____ **21.** Adding the letter "a" after a unit of pressure, as in psia, indicates ___ pressure.
                            A. about
                            B. atmospheric
                            C. approximate
                            D. none of the above

_____ **22.** ___ gauge pressure is gauge pressure less than zero.
                            A. Absolute
                            B. Negative
                            C. Positive
                            D. all of the above

Name _____   Date _____

Activity 9-1—Hydraulic Systems

A hydraulic system has a force applied to a small piston. Hydraulic liquid transmits the force to a large piston. The large piston applies the magnified force in the desired application. To increase the effectiveness of the hydraulic system, a mechanical arm can be added to multiply the force.

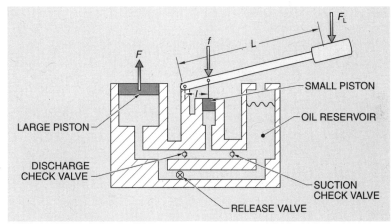

Calculate the forces as described in the questions.

_____ 1. What is the working force, F, in lb when the force applied to the small piston, f, is 20 lb? The area of the small piston, a, is 0.1963 sq in. and the area of the large piston, A, is 3.14 sq in.

_____ 2. What is the force applied to the small piston, f, in lb that is required to deliver a working force, F, of 1000 lb? The area of the small piston, a, is 0.1963 sq in. and the area of the large piston, A, is 6.28 sq in.

_____ 3. What is the working force, F, in lb when the force applied to the small piston, f, is 25 lb? The area of the small piston, a, is 0.0491 sq in. and the area of the large piston, A, is 12.56 sq in.

A handle is added to increase the force available to apply to the small piston. The length of the mechanical arm, L, is 18″ from the pivot point to the point where the force is applied. The length of the mechanical arm, l, is 1.25″ from the pivot point to the small piston.

_____ 4. What is the mechanical advantage of the lever arm?

_____ 5. What is the working force, F, in lb when the force applied to the mechanical arm, F_L, is 20 lb? The area of the small piston, a, is 0.1963 sq in. and the area of the large piston, A, is 3.14 sq in.

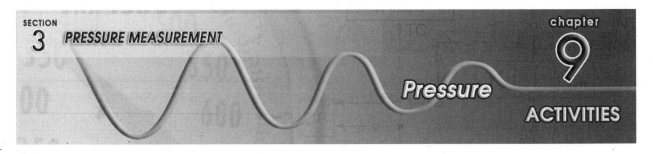

Name _____ **Date** _____

Activity 9-2—Hydrostatic Pressure

The pressure due to the height of a column of liquid depends on the height and density of the liquid. The storage tanks contain water.

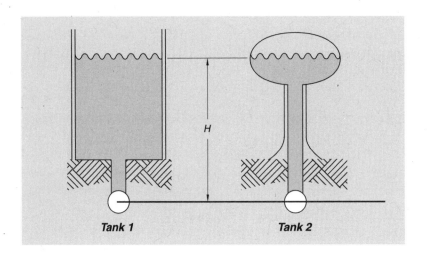

Tank 1   Tank 2

_____ **1.** What is the pressure in psig in the underground pipe for Tank 1 if H = 168′?

_____ **2.** What is the pressure in psig in the underground pipe for Tank 2 if H = 168′?

_____ **3.** What height of a column of mercury is equivalent to the 168′ of water?

Name _____ Date _____

Activity 9-3—Friction Losses

The hydraulic water system shown has a storage tank on top of a hill. The water flows downhill through a pipe to fill another storage tank. A pump is used to pump the water uphill to another storage tank. There are several places where there are pressure losses due to friction. The pump adds pressure back to the system to raise the level of the water in the highest tank. $H_1 = 102$ ft, $H_2 = 140$ ft, $H_3 = 135$ ft, and $\Delta P_3 = 10$ psi.

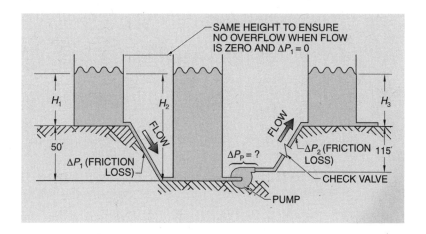

_____ **1.** What is the friction loss, ΔP_1, in ft of water head of the fluid flowing downhill through the pipe?

_____ **2.** What is the friction loss, ΔP_1, in psi of the fluid flowing downhill through the pipe?

_____ **3.** What is the differential pump pressure, ΔP_p, in ft of water head?

_____ **4.** What is the differential pump pressure, ΔP_p, in psi?

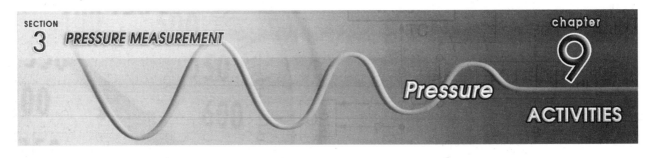

SECTION
3 PRESSURE MEASUREMENT

chapter
9

Pressure
ACTIVITIES

Name _____ **Date** _____

Activity 9-4—Absolute, Gauge, and Vacuum Pressures

All pressure measurements are differential pressure measurements. Gauge pressure is the pressure difference between a pressure measurement and the ambient atmospheric pressure. Absolute pressure is the difference between a pressure measurement and a perfect vacuum. Vacuum pressure is the difference between a pressure measurement and the ambient atmospheric pressure when the pressure measurement is less than the atmospheric pressure.

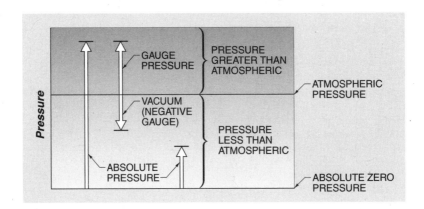

Convert the given pressures to the desired pressure units.

_____ **1.** What is the absolute pressure equivalent to 150.0 psig in psia?

_____ **2.** What is the absolute pressure equivalent to 4.3 psig in psia?

_____ **3.** What is the gauge pressure equivalent to 101.7 psia in psig?

_____ **4.** What is the gauge pressure equivalent to 90.0 psia in psig?

_____ **5.** What is the vacuum pressure equivalent to 4.3 psia in psi vacuum?

_____ **6.** What is the vacuum pressure equivalent to 14.7 psia in psi vacuum?

Name _____ Date _____

Activity 9-5—Pressure Conversions

Pressure is measured using many different units. It is often necessary to be able to convert between different units. For the following, convert the given pressure to the desired units. Not all units are given in the table.

	kg per sq cm	lb per sq in.	atm	bar	in. of Hg	kilopascals	in. of water	ft of water
kg per sq cm	1	14.22	0.9678	0.98067	28.96	98.067	394.05	32.84
lb per sq in.	0.07031	1	0.06804	0.06895	2.036	6.895	27.7	2.309
atm	1.0332	14.696	1	1.01325	29.92	101.325	407.14	33.93
bar	1.01972	14.5038	0.98692	1	29.53	100	402.156	33.513
in. of Hg	0.03453	0.4912	0.03342	0.033864	1	3.3864	13.61	11.134
kilopascals	0.0101972	0.145038	0.0098696	0.01	0.2953	1	4.02156	0.33513
in. of water	0.002538	0.0361	0.002456	0.00249	0.07349	0.249	1	0.0833
ft of water	0.03045	0.4332	0.02947	0.029839	0.8819	2.9839	12	1

* *NOTE*: Use multiplier at convergence of row and column

_____ 1. What is the equivalent of 2.1 psig when reported in kg/sq cm?

_____ 2. What is the equivalent of 2300 mm WC when reported in in. WC?

_____ 3. What is the equivalent of 85.3 in. WC when reported in psig?

_____ 4. What is the equivalent of 100.0 psig when reported in bar?

_____ 5. What is the equivalent of 75.2 psig when reported in kPa?

_____ 6. What is the equivalent of 7.13 kg/sq cm when reported in psig?

_____ 7. What is the equivalent of 15.0 in. Hg when reported in psig?

_____ **8.** What is the equivalent of 12.7 in. Hg when reported in in. WC?

_____ **9.** What is the equivalent of 1.50 psig when reported in in. WC?

_____ **10.** What is the equivalent of 23.3 kilopascals when reported in bar?

Name _____   **Date** _____

_____ **1.** A ___ is a device consisting of a liquid-filled tube used for measuring pressure.
    A. graduated cylinder
    B. manometer
    C. diaphragm
    D. transducer

T   F   **2.** Specific gravity is the ratio of volume of a fluid to volume of a reference fluid.

T   F   **3.** An inclined-tube manometer is a manometer with its measuring column at an angle to the horizontal to reduce the vertical height.

T   F   **4.** A transducer is a device that converts one form of energy to another, such as converting pressure to voltage.

T   F   **5.** A well-type manometer is a manometer with a vertical glass tube connected to a metal well, with the measuring liquid in the well at the same level as the zero point on the tube scale.

_____ **6.** A pressure spring is a mechanical pressure sensor consisting of a hollow tube formed into a ___ shape.
    A. C
    B. helical
    C. spiral
    D. all of the above

_____ **7.** A(n) ___ is a device consisting of metal, rubber, or plastic components such as diaphragms, capsules, springs, or bellows that flex in proportion to the pressure applied within or against them.

_____ **8.** A(n) ___ is a mechanical pressure sensor consisting of a one-piece, collapsible, seamless metallic unit with deep folds formed from thin-wall tubing.

T   F   **9.** A capsule is a mechanical pressure sensor consisting of two convoluted metal diaphragms with their outer edges welded together to provide an empty chamber between them.

_____ **10.** ___ is the ratio of the area of the tube to the area of the well in a well-type manometer.
   A. Area ratio
   B. Well drop
   C. Piezo ratio
   D. all of the above

_____ **11.** A ___ is a manometer used to measure atmospheric pressure.
   A. barometer
   B. transducer
   C. pressure spring
   D. capsule

_____ **12.** A(n) ___ is a mechanical pressure sensor that consists of a C-shaped tube flattened into an elliptical cross section.

Name _____   **Date** _____

Activity 10-1—Reading Manometers

The drawings show various U-tube manometers set up to measure air pressure. The size of the manometer and type of manometer fluid change from drawing to drawing.

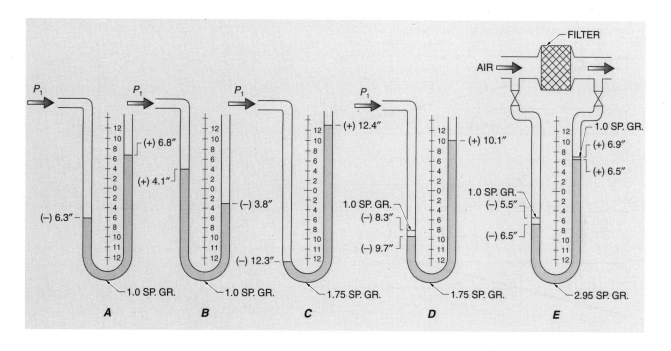

_____ **1.** What is the air pressure in in. WC being applied to manometer A?

_____ **2.** What is the air pressure in in. WC being applied to manometer B?

_____ **3.** What is the air pressure in in. WC being applied to manometer C?

_____ **4.** What is the air pressure in in. WC being applied to manometer D?

_____ **5.** What is the air pressure in in. WC being applied to manometer E?

Name _____   **Date** _____

Activity 10-2—Manometer Fluids

A 15″ U-tube manometer is to be used to measure the pressure drop across a water filter. The connecting lines and the manometer are filled with water. The manometer fluid must not be miscible with water. A clean filter has a pressure drop of 1.0 psid. A dirty filter has a pressure drop of 5.0 psid.

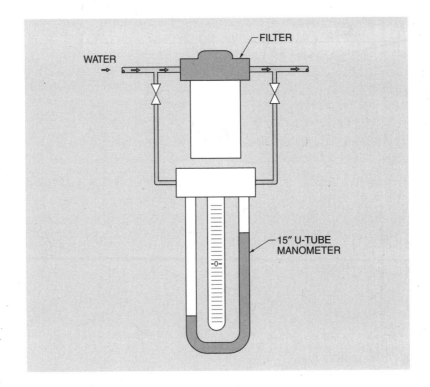

_____ 1. What is an appropriate manometer fluid for this application?

_____ 2. A manometer constant is the amount of pressure change per change in the height of the manometer fluid. What is a manometer constant to convert inches manometer fluid under water to psid?

SECTION
3 *PRESSURE MEASUREMENT*

chapter
11

Electrical Pressure Instruments

REVIEW
QUESTIONS

Name _____ **Date** _____

T   F   **1.** An electrical pressure transducer is a device that converts input energy from pressure into output electrical energy.

_____ **2.** ___ is the property of an electric circuit that opposes a magnetic flux.
       A. Piezoelectricity
       B. Capacitance
       C. Transduction
       D. Reluctance

_____ **3.** A(n) ___ pressure cell is a transducer that converts a differential pressure to an output signal.
       A. actuated
       B. super
       C. reluctance
       D. none of the above

T   F   **4.** A resistance pressure transducer is a mechanical transducer that measures the deformation of a rigid body as a result of the force applied to the body.

T   F   **5.** A pressure transmitter is a pressure transducer with a power supply and a device that conditions and converts the transducer output into a standard analog or digital output.

T   F   **6.** A strain gauge is a diaphragm pressure sensor with a capacitor as the electrical output element.

_____ **7.** A reluctance pressure transducer is a diaphragm pressure sensor with a metal ___ mounted between two stainless steel blocks.

T   F   **8.** A pressure transmitter is a pressure-sensing device that provides a discrete output (contact make or break) when applied pressure reaches a preset level within the switch.

_____ **9.** A ___ pressure transducer is a diaphragm pressure sensor with a capacitor as the electrical element.
       A. piezoelectric
       B. reluctance
       C. capacitance
       D. none of the above

_____ **10.** A(n) ___ pressure transducer is a diaphragm or bellows pressure sensor with electrical coils and a movable ferrite core as the electrical element.

SECTION
3 *PRESSURE MEASUREMENT*

chapter
12

Practical Pressure Measurement and Calibration

REVIEW
QUESTIONS

Name _____   **Date** _____

T   F   **1.** Dry legs can be used to measure the pressure of liquids in pipelines.

T   F   **2.** A three-valve manifold is unnecessary because the manometer can open up to the process stream without blowing out the manometer fluid.

T   F   **3.** A manometer is an impulse line that is filled with a noncondensing gas.

T   F   **4.** A wet leg is an impulse line filled with fluid that is compatible with the pressure-measuring device.

T   F   **5.** There can be moisture condensation in the connection piping or tubing between a process and a manometer.

_____ **6.** Pressure gauges must be protected from ___.
    A. pressure pulsations
    B. high temperatures
    C. corrosive materials
    D. all of the above

_____ **7.** ___ is subjecting a mechanical sensor to excessive pressure beyond the design limits of the instrument.
    A. Corrosion
    B. Steam condensation
    C. Overranging
    D. all of the above

T   F   **8.** When a U-tube manometer is used to measure pressure, a single-valve manifold is used with the manometer for easy shutoff and equalizing.

_____ **9.** When a pressure gauge on a boiler is mounted 8′ below the top of the water level of the wet leg, the pressure adjustment is ___ psi.
    A. 0
    B. 2.51
    C. 3.47
    D. 4.33

_____ **10.** ___ is the procedure for operating a manometer's three-way manifold valves to disconnect the manometer from a process.

_____ **11.** ___ cause pressure pulsations in a pipe line.
    A. Pumps starting and stopping
    B. Valves opening and closing
    C. Pipe vibrations
    D. all of the above

_____ **12.** A common difficulty with manometers that can cause errors in measurement is ___.
    A. leaking diaphragms
    B. varying capacitance
    C. moisture condensation
    D. all of the above

T   F   **13.** A technician should ensure that the proper calibration standard is used.

_____ **14.** Pressure gauges equipped with overpressure protection have a(n) ___ response to true pressure changes than pressure gauges without overpressure protection.
    A. faster
    B. slower
    C. equal
    D. all of the above

T   F   **15.** After the high-pressure valve on a three-way manifold is closed during the manometer disconnect process, a manometer can still contain pressurized fluids.

T   F   **16.** There is a single standard definition of the unit "inches of water."

T   F   **17.** Fluid can be blown out of a manometer by opening the connection to a process too quickly.

T   F   **18.** Pressure seals with integral sealing liquids should not be disassembled due to the difficulty of restoring that integrity.

T   F   **19.** High process temperatures can damage a pressure sensor.

T   F   **20.** Isolating sealing systems have been developed to protect pressure sensors from corrosive fluids.

_____ **21.** A(n) ___ is a hydraulic pressure-calibrating device that includes a manually operated screw press, a weight platform supported by a piston, a set of weights, and a fitting to connect the device to a gauge.

_____ **22.** Thermal expansion of a filling fluid due to ambient or process temperature changes has minimal effects on accuracy when the measurement range is ___ psig or more.
    A. 0
    B. 5
    C. 10
    D. 15

_____ **23.** Adding enough inlet tubing to allow the process fluid to cool before entering the sensor may ___ the useful temperature range of pressure sensors.
    A. decrease
    B. increase
    C. destroy
    D. not affect

Practical Pressure Measurement
and Calibration

Name _____ Date _____

Activity 12-1—Purging Systems

Instruments commonly need protection from corrosive or hot process fluids. Purging systems are a common way of protecting instruments. The fluid used to purge the sample lines must be compatible with the process fluid and with the instrument.

1. A low-pressure purging system is used to protect a draft pressure instrument from a condensate problem from contact with the wet process gas. Complete the drawing. Identify the composition and source of an appropriate purge fluid. Include any instruments required to measure and control the flow of the purge fluid.

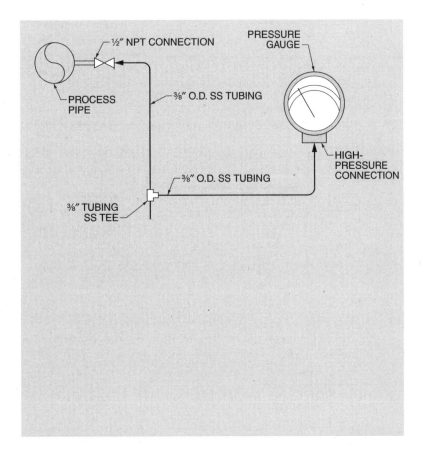

Name _____ **Date** _____

Activity 12-2—Elevation Changes

It is desired to measure a corrosive fluid in a process pipeline. Because of the corrosive fluid, the pressure transmitter needs to be protected with a chemical seal with a capillary connecting it to the transmitter. The pressure to be measured is 0 psig to 30 psig and the process connection is 15′ above the transmitter location. The capillary fluid has a specific gravity of 1.75. The transmitter was calibrated when the chemical seal was at the same elevation as the transmitter.

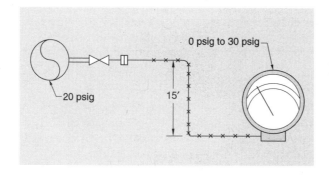

_____ **1.** What error in pressure measurement in psi would be sensed by the transmitter?

2. What simple changes can be made to allow the transmitter to read correctly?

_____

_____

_____

Differential Pressure Cells

There are many types of electrical transducers such as strain gauge, capacitance, inductance, reluctance, etc. These transducers are usually inside the electronics that comprise the modern pressure transmitter. The basic features displayed by all modern pressure transmitters were first developed for the d/p cell.

1. List and describe three technical innovations introduced by the d/p cell that revolutionized pressure measurement.

   _____

   _____

   _____

SECTION
4 LEVEL MEASUREMENT

chapter
13

Mechanical Level Instruments

REVIEW
QUESTIONS

Name _____ **Date** _____

T   F   **1.** Level measurement is essential for safety systems in boilers and for overflow and spill prevention systems in tanks and silos.

_____ **2.** Level is normally measured in linear units of ___ or translated into units of volume or weight.
    A. displacement
    B. length
    C. temperature
    D. all of the above

T   F   **3.** Armored gauge glasses are used in boilers and high-pressure vessels.

T   F   **4.** Standard d/p cells used to measure level are susceptible to attack by corrosive fluids or blockage by slurries.

_____ **5.** ___ level measurement is a method of measuring level where the concern is whether the level is within the desired limits.

_____ **6.** ___ level measurement is a method of measuring level over a range of values.
    A. Continuous
    B. Range
    C. Point
    D. Tracking

_____ **7.** A(n) ___ is a continuous level measuring instrument that consists of a glass tube connected above and below the liquid level in a tank and that allows the liquid level to be observed visually.

_____ **8.** A ___ is a continuous level measuring instrument consisting of a floating object connected by a chain, rope, or wire to a counterweight, which is the level pointer.
    A. displacer
    B. tape float
    C. bubbler
    D. d/p cell

_____ **9.** A(n) ___ glass has a vertical sawtooth surface that acts as a prism to improve readability.
    A. armored gauge
    B. gauge
    C. refracted gauge
    D. none of the above

_____ **10.** A cable and weight system consists of a ___.
    A. switch
    B. relay
    C. servomotor
    D. all of the above

T   F   **11.** An extended diaphragm d/p cell is a type of level measuring device that uses flanged diaphragm seals and capillary systems.

T   F   **12.** It is more convenient to measure pressure at the bottom of a tank than to measure the actual location of the top of the liquid in a sealed tank.

_____ **13.** ___ pressure variations present at the base of a liquid column provide the means of determining liquid level in a storage vessel.

T   F   **14.** As long as liquid in a tank has a constant density, variations in pressure are caused only by variations in level.

_____ **15.** Floats can be used to ___.
    A. actuate alarms
    B. indicate tank levels
    C. actuate switches
    D. all of the above

T   F   **16.** A paddle wheel switch is a continuous level measuring device consisting of a drive motor and a rotating paddle wheel mounted inside a tank.

T   F   **17.** A variation of the flush diaphragm d/p cell is the double-filled system, which is used with fluids containing high quantities of solids.

_____ **18.** A ___ is a liquid level measuring instrument consisting of a buoyant cylindrical object that is heavier than the liquid it is immersed in and is connected to a spring or torsion device that measures the buoyancy of the cylinder.
    A. displacer
    B. float
    C. tape float
    D. cable and weight

T   F   **19.** It is not possible to have condensed vapors collecting as liquids in low-pressure connections of d/p cell level measurement applications of pressurized tanks.

_____ **20.** A ___ is a level measuring instrument consisting of a tube extending to the bottom of a vessel; a pressure gauge, single-leg manometer, transmitter, or recorder; a flowmeter to adjust the flow rate of air or nitrogen through the tube; and a pressure regulator to limit the inlet pressure.
    A. float
    B. bubbler
    C. displacer
    D. cable and weight

_____ **21.** Level measurement is often used to measure the ___ of material in a tank or vessel.
    A. density
    B. temperature
    C. volume
    D. all of the above

Mechanical Level Instruments

Name _____   **Date** _____

Activity 13-1—Gauge Glasses

Armored flat gauge glasses are used as level indicators in high-pressure applications. The gauge glass is specified along with a pair of gauge valves to match the connection dimensions that exist on the pressure vessel. Specification sheets from the manufacturer are needed to specify the gauge glass and valves.

The flat gauge glass side connections are to be used with extended center sections. The side connection locations have the same spacing as the gauge overall length. Gauge valves with offsets and without offsets can be used. Valve orientations as shown in the illustration should be used. The offsets for the gauge values are 1¾″. Vessel connections are ¾″ NPT. It is best to select the gauge with the longest visible glass.

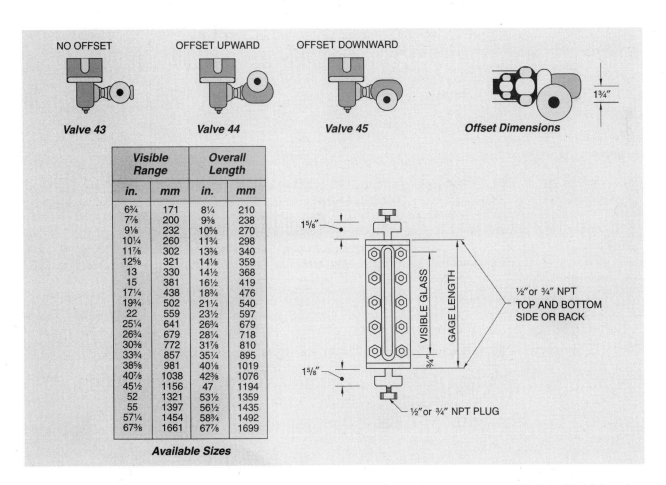

Visible Range		Overall Length	
in.	mm	in.	mm
6¾	171	8¼	210
7⅞	200	9⅜	238
9⅛	232	10⅝	270
10¼	260	11¾	298
11⅞	302	13⅜	340
12⅝	321	14⅛	359
13	330	14½	368
15	381	16½	419
17¼	438	18¾	476
19¾	502	21¼	540
22	559	23½	597
25¼	641	26¾	679
26¾	679	28¼	718
30⅜	772	31⅞	810
33¾	857	35¼	895
38⅝	981	40⅛	1019
40⅞	1038	42⅜	1076
45½	1156	47	1194
52	1321	53½	1359
55	1397	56½	1435
57¼	1454	58¾	1492
67⅜	1661	67⅞	1699

Available Sizes

For a connection dimension of 23½″, answer the following questions.

_____ **1.** What is the overall length?

_____ **2.** What top valve is required?

_____ **3.** What bottom valve is required?

For a connection dimension of 60″, answer the following questions.

_____ **4.** What is the overall length?

_____ **5.** What top valve is required?

_____ **6.** What bottom valve is required?

For a connection dimension of 50″, answer the following questions.

_____ **7.** What is the overall length?

_____ **8.** What top valve is required?

_____ **9.** What bottom valve is required?

Name _____   Date _____

Activity 13-2—Bubblers

A bubbler system is installed in the top of a vessel for measuring the level of the liquid in the vessel. The vessel contains a liquid with a specific gravity of 1.1. A single-leg manometer is used to measure the pressure. Single-leg manometer sizes of 10″, 15″, and 20″, and manometer fluids with specific gravities of 1.0, 1.5, 1.75, 2.95, and mercury with a specific gravity of 13.6 are available.

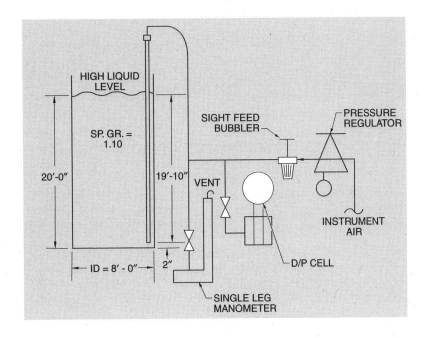

_____   **1.** What is the maximum pressure to be measured, in in. WC?

2. Select the best combination of manometer and fluid for this application.

_____

_____

_____

_____

_____ **3.** A manometer can be scaled to show the actual level corresponding to the movement of the manometer fluid in inches of manometer fluid per foot of liquid in the vessel. Design a scale for the selected single leg manometer and fluid in Question 2 to represent the vessel level in feet and tenths of feet.

_____ **4.** A d/p cell is added to the measurement system so that the value can be transmitted to the control room. Calculate the d/p cell calibration range.

SECTION
4 LEVEL MEASUREMENT

chapter
14

Electrical Level Instruments

REVIEW
QUESTIONS

Name _____   **Date** _____

_____ 1. A(n) ___ is a point level measuring system consisting of a circuit of two or more probes or electrodes, or an electrode and the vessel wall where the material in the vessel completes the circuit as the level rises in the vessel.
    A. capacitance sensor
    B. conductivity probe
    C. electrode sensor
    D. magnetostrictive probe

_____ 2. A ___ is an insertable device that contains a sensor.
    A. dielectric
    B. thermocouple
    C. capacitor
    D. probe

T   F 3. For RF capacitance continuous level measuring instruments, a vertically suspended capacitance probe can be used for measuring the level in a tank.

_____ 4. A(n) ___ sensor is a point level measuring instrument consisting of a light source and a detector that measures the level of the contents of a vessel when the beam is broken.

T   F 5. RF capacitance probe level detectors can operate with granular solids as well as liquids.

T   F 6. Conductivity level probes can only be used with conductive liquids.

_____ 7. A(n) ___ sensor is a liquid point level measuring instrument that is based on the same principles as the reflex gauge glass.

_____ 8. A ___ is any device that is sensitive to a change in the measured phenomenon or characteristic.
    A. resistor
    B. diaphragm
    C. sensor
    D. capacitor

_____ 9. A magnetostrictive sensor is a part of a continuous level measuring system consisting of a(n) ___.
    A. electronics module
    B. waveguide
    C. float containing a magnet
    D. all of the above

T   F   **10.** Magnetostrictive refers to the ability of an electrical device to store charge as the result of the separation of charge.

_____ **11.** ___ is the ability of a circuit to conduct alternating current and is the reciprocal of impedance.
     A. Admittance
     B. Inductance
     C. Resistance
     D. Reactance

T   F   **12.** Probes may be supplied with low-level AC or high-level DC voltage, depending on the nature of the process liquid.

_____ **13.** An inductive probe is a point level measuring instrument consisting of a sealed probe containing a ___.
     A. resistor
     B. coil
     C. capacitor
     D. all of the above

T   F   **14.** Capacitance is the insulating material between the conductors of a capacitor.

_____ **15.** The amount of capacitance depends on the ___.
     A. dielectric constant
     B. surface area of the conductors
     C. distance between the conductors
     D. all of the above

_____ **16.** Common types of electrical level measuring systems include ___.
     A. conductivity probes
     B. magnetostrictive sensors
     C. thermal dispersion sensors
     D. all of the above

_____ **17.** A(n) ___ is a part of a level measuring instrument and consists of a metal rod inserted into a tank or vessel, with a high-frequency alternating voltage applied to it and a means to measure the current that flows between the rod and a second conductor.

_____ **18.** A(n) ___ sensor is a point level measuring instrument consisting of two probes that extend from the detector into a vessel, with one of the probe tips being heated.

T   F   **19.** Dielectric refers to a property of certain ferrous alloys having dimensions that change in response to magnetic stress.

T   F   **20.** Capacitance is the ratio of the insulating ability of a material to the insulating ability of a vacuum.

T   F   **21.** The buildup of process material on the probe sometimes presents a problem because the presence of a coating can act as a dielectric even when the level is below the probe.

T   F   **22.** A thermal dispersion sensor's detector monitors the difference in temperature between a heated probe tip and an unheated probe.

_____ **23.** A(n) ___ is an electrical device made up of two conductors separated by an insulating material.

Name _____ Date _____

Activity 14-1—Induction Relays

Conductive probe level instruments are commonly available with induction relays. An induction relay depends on the presence of enough voltage to force sufficient current through the process liquid to energize the relay. Induction relays consist of primary and secondary coils. The primary coil voltage must be chosen to match the line voltage of the control system. The choice of secondary coil depends on the conductivity of the process liquid. A low-voltage coil can be used for very conductive liquids and a high-voltage coil needs to be used for liquids with low conductivity.

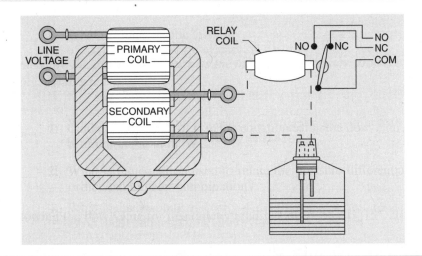

Recommended Secondary Coils for Various Types of Liquids

Secondary Coil Voltage	Typical Liquids	Max. Specific Resistance of Liquid (ohm–cm)	Min. Specific Conductivity of Liquid (µS/cm)
12	Metallic circuits.	15	67,000
24	Metallic circuits.	61	16,000
40	Acid or caustic solutions; milk; brine and salt solutions; plating solutions; buttermilk; soups.	216	4,630
90	Weak acid or caustic solutions; beer; baby foods; fruit juices.	1,065	940
220	Sewage; most water - except very soft; pottery slip; water soluble oil solutions; starch solutions.	6,650	150
360	Very soft water; sugar syrup.	17,000	59
480	Steam condensate; strong alcohol solutions.	26,600	38
800	Demineralized or distilled water.	92,600	11

Use the table to specify the lowest voltage that will work for the specified liquid. The specific conductivity is given in each problem.

_____ **1.** Sea water, 30,000 μS

_____ **2.** Boiler water, 10 μS

_____ **3.** Good quality raw water, 60 μS

_____ **4.** Steam condensate, 38 μS

5. What safety problems are present when using the induction relay level sensors for liquids with low conductivity?

_____

_____

Name _____   Date _____

Activity 14-2—Condensate Level Control

A vertical tank with dished heads is used to collect condensate from various users. The condensate is then pumped to a deaeration unit before being reused. It is desired to use a solid-state relay conductive probe system to provide a high and low alarm. The following information is necessary to select all the parts needed.

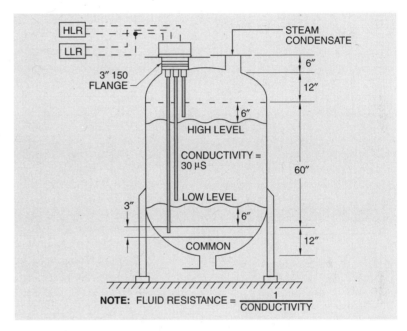

SENSITIVITY SELECTION

T/W level control systems use the liquid as an electrical conductor to complete the Series 25 relay sensing circuit, and it is necessary that the relay have an operating sensitivity greater than the resistance of the liquid to be controlled. The sensitivity of the Series 25 relay is determined by the value of the resistor used. When operating from contacts of pilot switches, any of the resistors can be used, but the smallest resistor value is recommended. The proper resistor must be selected during installation.

High Sensitivity Relay	Nominal Resistance	Part Number
Maximum Sensitivity	10,000 Ohms	05 - 1000
Direct Operation: 11.6 Megohms	22,000 Ohms	05 - 2200
Inverse Operation: 12.0 Megohms	68,000 Ohms	05 - 6800
	0.33 Megohms	05 - 3300
Electrode Potential	0.82 Megohms	05 - 820
9.6 volts DC	2.2 Megohms	05 - 2200
	5.6 Megohms	05 - 5600
Electrode Current	12.0 Megohms	05 - 1200
Less than 1 Milliampere	Variable	05 - x100
	Variable	05 - x200
	Variable	05 - x300

FLANGED CAST IRON ELECTRODE HOLDERS

The flanged holders are for use with T/W wire suspension electrodes in nonpressure applications such as elevated tanks, water reservoirs, underground storage tanks, sewage wet wells, and other installations requiring relatively long electrode lengths. Flange holders are available in 3″, 4″, and 6″ sizes that fit the available 125 lb cast iron and 150 lb steel mounting flange specifications. These holders have a removable inner plate provided with a grommet for each wire suspension electrode. The grommets are sized to give a snug, vapor-tight fit with T/W type SW electrode suspension wire.

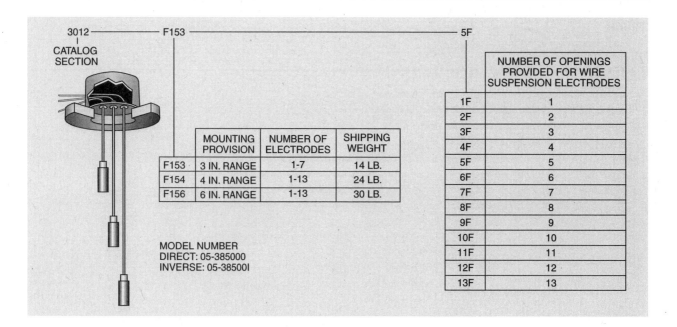

MOUNTING PROVISION	NUMBER OF ELECTRODES	SHIPPING WEIGHT
F153	3 IN. RANGE	1-7	14 LB.
F154	4 IN. RANGE	1-13	24 LB.
F156	6 IN. RANGE	1-13	30 LB.

MODEL NUMBER
DIRECT: 05-385000
INVERSE: 05-38500I

	NUMBER OF OPENINGS PROVIDED FOR WIRE SUSPENSION ELECTRODES
1F	1
2F	2
3F	3
4F	4
5F	5
6F	6
7F	7
8F	8
9F	9
10F	10
11F	11
12F	12
13F	13

ELECTRODES

Type 1-032 is intended for corrosive liquids. It is 3″ long, made from a ½″ diameter rod, and available in any of the materials listed in the table. The type SW wire is attached at the factory and the connection is completely sealed with PVC and bonded to the insulation of the SW wire.

_____ **1.** What is the fluid resistance, in ohms?

_____ **2.** What is the part number of the required high-level relay? Action is to be direct.

_____ **3.** What is the required resistor for the high-level relay?

_____ **4.** What is the part number of the required low-level relay? Action is to be inverse.

_____ **5.** What is the required resistor for the low-level relay?

_____ **6.** Determine the required cast iron electrode holder. List the catalog number.

7. List the lengths of the three suspension electrodes that are needed.

_____

_____

_____

SECTION
4 *LEVEL MEASUREMENT*

chapter
15

Ultrasonic, Radar, and
Laser Level Instruments

REVIEW
QUESTIONS

Name _____ **Date** _____

_____ **1.** A ___ is another name for guided wave radar.
  A. magnetostrictive sensor
  B. pulsed radar
  C. frequency modulated continuous wave
  D. time domain reflectometer

_____ **2.** A ___ sensor is a level measuring sensor consisting of a radar generator that directs an intermittent pulse with a constant frequency toward the surface of the material in a vessel.
  A. tuning fork
  B. frequency modulated continuous wave
  C. pulsed radar level
  D. guided wave radar

_____ **3.** Transit time depends on the ___ of the signal in the air or vapor space above the material.

_____ **4.** ___ is the time it takes for a transmitted ultrasonic signal to travel from the ultrasonic level transmitter to the surface of the material to be measured and back to the receiver.

T   F   **5.** For ultrasonic point level measurement, a rising liquid fills the gap between two crystals and blocks the signal.

_____ **6.** ___ affects the speed of the ultrasonic signal.
  A. Temperature
  B. Pressure
  C. Humidity
  D. all of the above

T   F   **7.** A tuning fork level detector is a point level measuring instrument consisting of a vibrating wire that resonates at a particular frequency and the circuitry to measure that frequency.

_____ **8.** A(n) ___ radar is a level measuring sensor consisting of an oscillator that emits a continuous microwave signal that repeatedly varies its frequency between a minimum and maximum value.

_____ **9.** The speed of a radar signal is equal to the speed of ___.
  A. sound in a vacuum
  B. sound in air
  C. light in air
  D. none of the above

_____ **10.** A laser level instrument is a level measuring instrument consisting of a ___.
    A. laser beam generator
    B. timer
    C. detector
    D. all of the above

_____ **11.** Ultrasonic level measurement cannot be used ___ or with liquids that are covered with foam.
    A. indoors
    B. in a vacuum
    C. outdoors
    D. with granular solids

_____ **12.** ___ is the range of frequencies from the minimum to the maximum value in an FMCW radar level sensor.
    A. Bandwidth
    B. Transit time
    C. Sweeptime
    D. Time domain reflectometry

T   F **13.** The electronic circuitry in an ultrasonic receiver measures the transit time and calculates the distance from the receiver to the surface of material.

_____ **14.** ___ is the constant time for an FMCW emitter to vary the frequency from the lowest frequency to the highest.

T   F **15.** Laser beams are intense, narrow light beams that can travel long distances.

_____ **16.** A ___ is a level measuring detector consisting of a cable or rod as the wave carrier extending from the emitter down to the bottom of the vessel and electronics to measure the transit time.
    A. laser level instrument
    B. tuning fork
    C. guided wave radar
    D. capacitance probe

_____ **17.** A(n) ___ is a level measuring instrument consisting of two electrically energized crystals, with one crystal used as a transmitter that generates a high-frequency sound and the other used as a receiver.

T   F **18.** Mixer blades and other objects in the vessel can cause false echoes, and dust in the air above a granular solids surface may absorb the radar signal.

_____ **19.** The ___ of a material determines the degree of absorption and, therefore, the strength of a reflected radar wave.
    A. viscosity
    B. dielectric constant
    C. density
    D. refractive index

Name _____   Date _____

Activity 15-1—Radar Instruments

It is desired to measure the level of 32% hydrochloric acid (HCl) in a fiberglass vessel using a radar level instrument. The radar instrument is supplied with 115 VAC power. The output signal is to be powered by the receiving instrument. The corrosive nature of the fluid requires the use of a PTFE window and SS spool piece. A nozzle is a short stub of piping connected to a vessel or tank during construction. Nozzles provide attachment points for piping, safety valves, and instruments to be connected to the vessel.

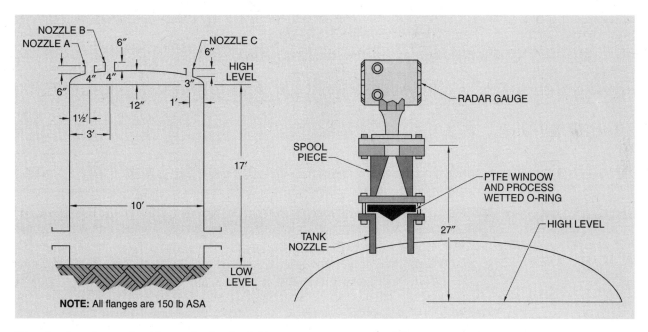

NOTE: All flanges are 150 lb ASA

Use the drawing and table on the back of this page to answer the following question.

1. Given the dimensions of the HCl vessel, select the most desirable nozzle on the vessel to mount the radar level instrument.

_____

_____

_____

_____

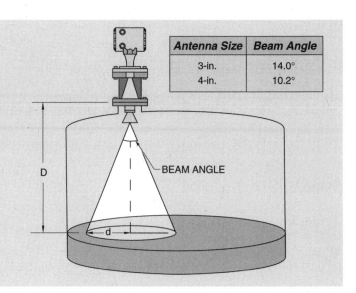

Distance D	Distance d	
From Gauge	3-in. Antenna	4-in. Antenna
ft	ft	ft
2	0.2	0.2
4	0.5	0.4
5	0.6	0.4
10	1.2	0.9
15	1.9	1.3
20	2.5	1.8
25	3.1	2.0

Antenna Size	Beam Angle
3-in.	14.0°
4-in.	10.2°

Consider the following factors when selecting an antenna:

• A 4-inch antenna is suggested for all installations because it produces a stronger signal. Only select a 3-inch antenna if there are physical limitations of the tank opening.

• Verify that *most* of the beam is within the tank walls.

• Avoid mounting the gauge over obstacles within the tank.

Use the table below to complete the following activity.

2. Select the required instrument by developing the model number.

_____

_____

_____

Model	Product Description
RDR	Radar Level Gauge

Code	Power Supply
AC	90–250 Volts AC
DC	18–36 Volts DC

Code	Material of Construction
CA	Carpenter 20
SS	316L Stainless Steel
TI	Titanium

Code	Antenna Size
3	Fits 3-in. opening
4	Fits 4-in. opening

Code	Mounting Flange Size
03	3-in. ASME B 16.5 (ANSI) (DN 80)
04	4-in. ASME B 16.5 (ANSI) (DN 100)
06	6-in. ASME B 16.5 (ANSI) (DN 150)

Code	Mounting Flange Rating
B1	ASME B 16.5 (ANSI) Class 150
B3	ASME B 16.5 (ANSI) Class 300
B6	ASME B 16.5 (ANSI) Class 600
D2	DIN PN 16
D4	DIN PN 40
D6	DIN PN 64

Typical Model Number
RDR A CA 3 03 B1

Spool Pieces

Flange Rating	Material	DN 80	DN 100	3-in. Part No.	4-in. Part No.
ASME B 16.5 (ANSI) Class 150	304 SS	NA	NA	331-081S	332-081S
	CS Painted	NA	NA	331-081C	332-081C
DIN PN 16	304 SS	321-081S	322-081S	NA	NA
	CS Painted	321-081C	322-081C	NA	NA

Bolt Kits

Material	3-in. Part No.	4-in. Part No.
Class 150 SS per ASTM F593	015-100S	016-100S
Class 150 CS ASTM A193, A194	015-100C	016-100C

Name _____   Date _____

_____ 1. ___ load cells are used where a tank hangs from a ceiling or beam.
A. Compression-type electronic
B. Tension-type
C. Hydraulic-type
D. Shear-type

T   F   2. Strain gauge load cells are generally in the form of a beam, column, or other stress member with strain gauges bonded to them.

T   F   3. A system is calibrated when the amount of applied weight over the entire usable range matches the scale readout within the error specified by the manufacturer.

_____ 4. The three forms of electronic load cells are the ___ types.
A. shear, expansion, and tension
B. compression, shear, and expansion
C. shear, compression, and tension
D. compression, expansion, and tension

T   F   5. Hydraulic load cells are more susceptible to errors caused by piping stresses than electronic load cell systems.

_____ 6. ___ load cells are used for long horizontal vessels where one end of the vessel needs to be free floating to allow for dimensional changes with temperature changes.

_____ 7. Load cells are selected with a safety margin that is ___% to ___% greater than the maximum calculated load that is expected to be applied to each load cell.
A. 0; 50
B. 0; 100
C. 50; 100
D. 100; 200

T   F   8. A significant difficulty with the use of load cells is that piping restraints tend to prevent movement of the vessel.

_____ 9. A piston compresses fluid in a hydraulic load cell that results in a change in ___ proportional to the load.

_____ 10. The tare weight is the weight of the ___ that is supported by the load cell(s).
A. piping
B. equipment
C. vessel
D. all of the above

_____ **11.** Electronic load cells only compress approximately ___″ when fully loaded.
      A. 0
      B. 0.005
      C. 0.050
      D. 0.500

_____ **12.** ___ are either piston-cylinder devices that produce a hydraulic output pressure or strain gauge assemblies that provide electrical output proportional to an applied load.

_____ **13.** For a nuclear level instrument, elements such as cesium 137 or cobalt 60 provide a(n) ___ in the form of gamma rays.
      A. ultrasonic frequency
      B. radioactive source
      C. radar signal
      D. transit time

T   F   **14.** The greater the vertical movement, the smaller the effect of the piping restraint.

T   F   **15.** All load cell systems require some vertical movement with increased load.

T   F   **16.** Hydraulic load cells are part of an open hydraulic pressure system in which the load cell transfers the pressure acting on the cell from the weight of the vessel and its contents to another load cell.

SECTION
4 LEVEL MEASUREMENT

chapter
16
ACTIVITIES

Nuclear Level Instruments
and Weigh Systems

Name _____ **Date** _____

Activity 16-1—Load Cells

A vertical vessel with a nominal capacity of 5000 gal. is to be mounted on four electronic shear-type load cells. The system is designed so that in an extreme case, three of the four load cells can carry the entire weight. The liquid specific gravity is 1.15. The tare weight of the empty vessel is 5380 lb. Shear-type load cells are available in the following ranges: 2000 lb, 5000 lb, 10,000 lb, 15,000 lb, 20,000 lb, and 30,000 lb.

1. Select a measurement range for the weigh system.

_____
_____
_____

2. Using the measured weight and the tare weight, select a set of four electronic shear-type load cells for this application.

_____
_____
_____

———————————— **3.** The electronic load cell readout has four active digits to display the weight. What is the smallest weight that can be displayed?

———————————— **4.** Given an accuracy of 0.1%, what is the maximum inaccuracy in lb when the tank is full?

———————————— **5.** Given an accuracy of 0.1%, what is the maximum inaccuracy in lb when the tank is 10% full?

Name _____ **Date** _____

_____ 1. ___ is a condition arising in a silo when material has built up over the feeder, blocking all flow out of the silo.
   A. Bridging
   B. Jamming
   C. Ratholing
   D. Mass flow

T   F   2. A bulk solid is a granular solid, such as gravel, sand, sugar, grain, cement, or other solid material, that can be made to flow.

_____ 3. ___ is a condition arising in a silo when material has built up over the feeder, blocking all flow out of the silo.

_____ 4. ___ flow is where material empties out of the bottom of a silo and the main material flow is down the center of the silo, with stagnant areas at the sides and bottom of the silo.
   A. Rathole
   B. Funnel
   C. Mass
   D. Bridging

_____ 5. A(n) ___ provides a signal that simulates the signal from load cells at various weights.
   A. electronic readout
   B. tare weight
   C. bridge
   D. load cell simulator

_____ 6. A(n) ___ is a boiler fitting that shuts the burner OFF in the event of a low water condition.

_____ 7. ___ flow is where all material in a silo flows down toward the bottom at the same rate.

_____ 8. ___ is a condition arising in a silo when material in the center has flowed out the feeder at the bottom, leaving large areas of stagnant material on the sides.

_____ 9. A transmitter can be protected from a corrosive process fluid by connecting capillary tubing, with a(n) ___, that is filled with liquid of constant specific gravity.

T   F   10. A water column is a boiler fitting that reduces the movement of boiler water to provide an accurate water level in the gauge glass.

_____ **11.** ___ are valves located on a water column that are used to determine boiler water level if a gauge glass is not functional.
    A. Control valves
    B. Gauge glasses
    C. Water columns
    D. Try cocks

T   F   **12.** Ratholing is the process of discharging water and undesirable accumulated material from a boiler.

T   F   **13.** A loss of water in a boiler can lead to the burning out of tubes and/or a boiler explosion.

_____ **14.** A(n) ___ is a granular solid that can be made to flow.

Name _____ Date _____

Activity 17-1—Load Cell Calibration

Weigh systems are typically designed to have accuracies of 0.1% of the measured weight. A common method of calibrating weigh systems is to apply known physical weights to the vessel and compare the known weights to the measured weight. The difference is plotted on a funnel chart, which shows the (+) or (−) error in pounds graphed against the applied weight. The 0.1% limit lines are drawn on the curve and form the funnel shape.

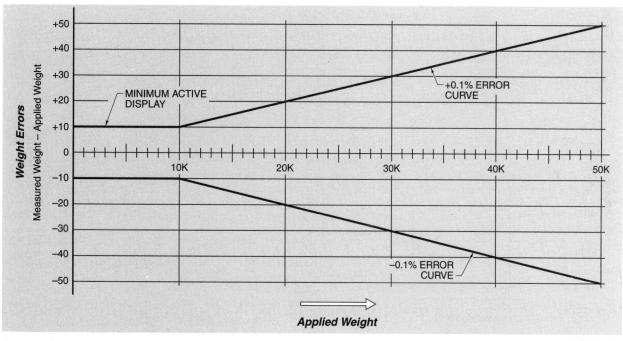

A record of the applied weight and the measured weight is shown below. Use this information to complete the activity.

Actual Applied Weight	Measured Weight
0	0
5000	5000
10,000	10,010
15,000	15,010
20,000	20,020
25,000	25,030
30,000	30,040
35,000	35,050
40,000	40,050
45,000	45,050

1. Plot the weight error against the applied weight on the funnel chart.

_____ 2. Did the test show that the weigh system was in calibration?

Name _____   Date _____

Activity 17-2—Load Cell Corrections

A weigh vessel was found to be out of calibration. After correcting some piping problems, a second calibration test was run, giving the following data. Use the data to complete the activity.

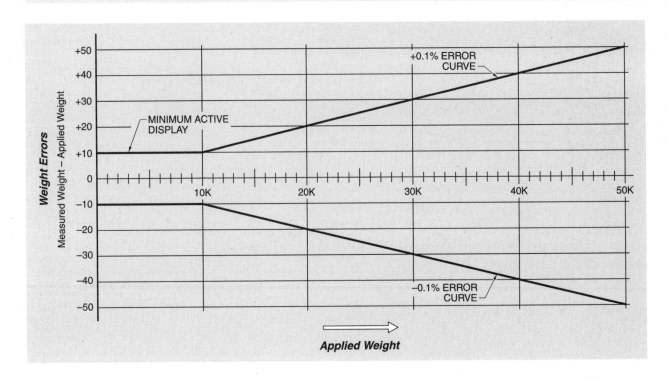

Actual Applied Weight	Measured Weight
0	0
5000	5000
10,000	10,010
15,000	15,010
20,000	20,000
25,000	24,990
30,000	29,980
35,000	34,980
40,000	39,970
45,000	44,970

1. Plot the error against applied weight on the funnel chart.

_____   **2.** Did the test show that the weigh system was in calibration?

Practical Level Measurement and Calibration
ACTIVITIES

Name _____   Date _____

Activity 17-3—Differential Pressure Cell Calibration

A pressurized vessel is filled with liquid and the pressure is measured with a d/p cell. The low-pressure side of the cell is connected to a purge system. The d/p cell needs to be calibrated. The calibration is done by determining the pressure difference between the high side and low side of the d/p cell in inches WC.

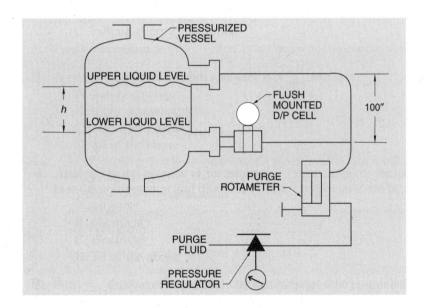

The tank contains water with a specific gravity of 1.0. The maximum height of the liquid in the vessel, h, is 86". The purge system is turned OFF so that the connection tube to the low-pressure side of the d/p cell is empty and the cell is directly exposed to the pressure in the tank above the water.

_____   **1.** Calculate the d/p cell calibration range.

The tank contains a liquid with a specific gravity of 1.1. The maximum height of the liquid in the vessel, h, is 96". The connection tube on the low-pressure side of the d/p cell is purged with nitrogen.

_____   **2.** Calculate the d/p cell calibration range.

The tank contains a liquid with a specific gravity of 1.1. The maximum height of the liquid in the vessel, h, is 96″. The connection tube on the low-pressure side of the d/p cell is purged with water.

_____ **3.** Calculate the d/p cell calibration range.

Name _____ **Date** _____

Level Scale for Weight

Level measurements can also be represented as weight as long as the specific gravity is known. The tank is 8′-0″ in internal diameter with a 20′-0″ maximum height. A single-leg manometer is filled with mercury.

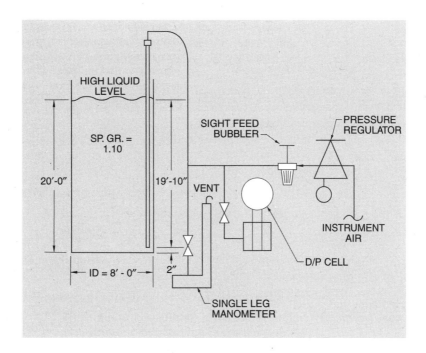

1. Using the tank as illustrated, develop a scale in pounds of material per inch of mercury movement for the single-leg manometer.

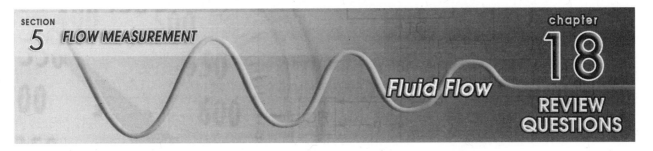

SECTION
5 **FLOW MEASUREMENT**

Fluid Flow

chapter
18
REVIEW
QUESTIONS

Name _____ **Date** _____

_____ 1. A streamline is a line that shows the direction and ___ of smooth flow at every point across a pipe profile.
   A. area
   B. temperature
   C. magnitude
   D. velocity

_____ 2. ___ is mass per unit volume.

_____ 3. ___ law is a gas law that states that the absolute pressure of a given quantity of gas varies inversely with its volume provided the temperature remains constant.

_____ 4. ___ is the quantity of fluid passing a point at a particular moment.
   A. Total flow
   B. Flow rate
   C. Absolute viscosity
   D. Kinematic viscosity

_____ 5. ___ is a ratio of the density of a fluid to the density of a reference fluid.
   A. Reynolds number
   B. Absolute viscosity
   C. Specific gravity
   D. Kinematic viscosity

_____ 6. A(n) ___ fluid is a fluid where there is very little change in volume when the fluid is subjected to a change in pressure.
   A. compressible
   B. incompressible
   C. kinematic
   D. laminar

_____ 7. The ___ number is the ratio between the inertial forces moving a fluid and viscous forces resisting that movement, and describes the nature of the fluid flow.

     T   F   8. The type of fluid is the most important characteristic of a fluid that affects flow.

_____ 9. A(n) ___ is a material that flows and takes the shape of its container.

_____ 10. ___ viscosity is the resistance to flow of a fluid and has units of centipoise (cp).
   A. Absolute
   B. Fluid
   C. Kinematic
   D. all of the above

91

_____ 11. ___ viscosity is the ratio of absolute viscosity to fluid density and has units of centistokes (cS).

T   F   12. A flow profile is a representation of the velocity of a fluid at different points across a pipe or duct.

T   F   13. A compressible fluid is a fluid where the volume and density change when the fluid is subjected to a change in pressure.

_____ 14. ___ flow is smooth fluid flow that has a flow profile that is parabolic in shape, and there is no mixing between the streamlines.

_____ 15. ___ law is a gas law that states that the volume of a given quantity of gas varies directly with its absolute temperature provided the pressure remains constant.

T   F   16. Smooth flow is fluid flow in which the flow profile is a flattened parabola, the streamlines are not present, and the fluid is freely intermixing.

_____ 17. If the absolute pressure on a gas is doubled while the temperature is held constant, the volume of the gas is ___.
   A. doubled
   B. halved
   C. not changed
   D. cannot be determined

T   F   18. Density is the pressure and temperature of a gas or vapor at the point of measurement.

_____ 19. ___ law is a gas law that states that the absolute pressure of a given quantity of a gas varies directly with its absolute temperature provided the volume remains constant.

T   F   20. Standard conditions are an accepted set of temperature and pressure conditions used as a basis for measurement.

_____ 21. ___ is the quantity of fluid that passes a point during a specific time interval.
   A. Viscosity
   B. Laminar flow
   C. Turbulent flow
   D. Total flow

Name _____   Date _____

Activity 18-1—Flow Rate Conversions

Flow is measured using many different units. It is often necessary to be able to convert between different units. For the following, convert the given flow to the desired units. Not all units are given in the table.

Given Value	To Convert to					
	gpm	gph	l/min	m³/hr	cm³/min	ft³/min
gpm	1	gpm × 60	gpm × 3.785	gpm × 0.2271	gpm × 3785	gpm × 0.1337
gph	gph × 0.01667	1	gph × 0.06309	gph × 0.003785	gph × 63.09	gph × 8.022
l/min	l/min × 0.2642	l/min × 15.85	1	l/min × 0.06	l/min × 1000	l/min × 0.0353
m³/hr	m³/hr × 4.403	m³/hr × 264.2	m³/hr × 16.67	1	m³/hr × 16,667	m³/hr × 0.5886
cm³/min	cm³/min × 0.0002642	cm³/min × 0.01585	cm³/min × 0.001	cm³/min × 0.00006	1	cm³/min × 0.0000353
ft³/min	ft³/min × 7.479	ft³/min × 448.7	ft³/min × 28.31	ft³/min × 1.699	ft³/min × 28,312	1

_____ 1. What is the equivalent of 22 gpm when reported in gph?

_____ 2. What is the equivalent of 5.2 gpm when reported in l/min?

_____ 3. What is the equivalent of 10.1 m³/hr when reported in gpm?

_____ 4. What is the equivalent of 5027 cm³/min when reported in gpm?

_____ 5. What is the equivalent of 4.78 gpm when reported in ft³/min?

_____ 6. What is the equivalent of 0.00348 m³/hr when reported in cm³/min?

_____ 7. What is the equivalent of 19.48 l/min when reported in ft³/min?

_____ 8. What is the equivalent of 8920 cm³/min when reported in ft³/min?

_____ 9. What is the equivalent of 15.3 ft³/min when reported in ft³/hr?

_____ 10. What is the equivalent of 3.168 gpm when reported in ft³/hr?

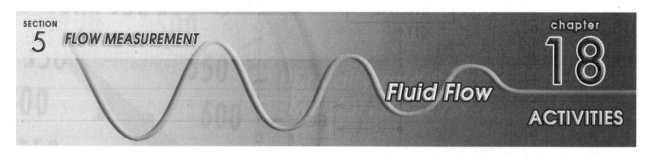

Name _____ Date _____

Activity 18-2—Volumetric Flow

An aboveground swimming pool installation has just been completed. The pool is 24′ in diameter and needs to be filled to the middle of the skimmer overflow outlet. It is desired to fill the pool with a garden hose.

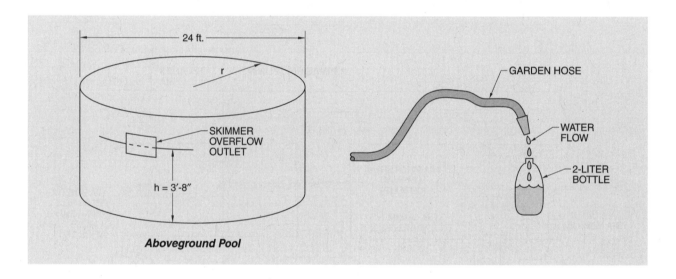

Aboveground Pool

_____ 1. What is the volume of water in gallons that is needed to fill the pool?

_____ 2. How long does it take to fill the pool if it takes 10 sec to fill a 2-liter bottle?

Name _____   **Date** _____

Activity 18-3—Reynolds Number

It is desired to pipe water from one location in a plant to another. There is an existing unused 3″ Schedule 40 pipe already installed between the two locations. A flowmeter is available that requires that the flow be in the turbulent range. Water density is 62.4 lb/ft³ and water viscosity is 1.0 cp.

_____   **1.** What is the Reynolds number if the maximum flow is 20 gpm?

_____   **2.** Is the flow turbulent?

_____   **3.** What is the lowest flow in gpm that can be measured if the lowest turbulent flow Reynolds number is 4000?

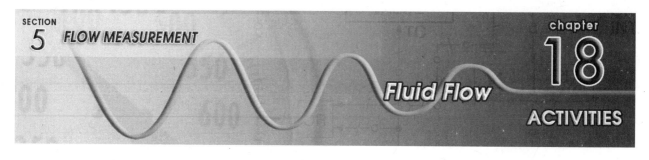
Name _____   Date _____

Activity 18-4—Gas Laws

Air or gas flow is often given in acfh, actual cubic feet per hour, or scfh, standard cubic feet per hour. The unit acfh is used for actual volumetric flow rate at the given pressure and temperature; scfh is the same mass of gas, but given as if the mass of air were at a standard pressure of 14.7 psia and 70°F. The gas laws are used to convert from one flow rate to another.

An air compressor has an atmospheric inlet pressure. The air is compressed, cooled, and fed into an air receiver. Air from the receiver passes through an orifice plate flowmeter. The pressure in the receiver is 83 psig and the temperature is 112°F. The design flow at the location of the flowmeter is 1206 acfh at 83 psig and 112°F.

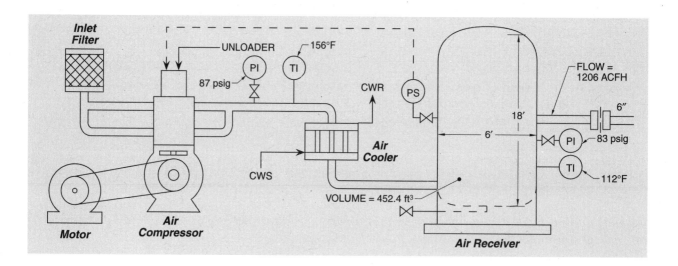

_____   1. What is the design flow of air at scfh conditions of 14.7 psia and 70°F used for the flowmeter using the combined gas law?

_____   2. What is the actual flow rate in acfh at the discharge of the compressor before the cooler?

Name _____ **Date** _____

Activity 18-5—Compressed Air

An air compressor has an atmospheric inlet pressure. The air is compressed, cooled, and fed into an air receiver. Air from the receiver passes through an orifice plate flowmeter. The pressure in the receiver is 83 psig and the temperature is 112°F. The compressor shuts off and the air in the receiver cools off to 70°F.

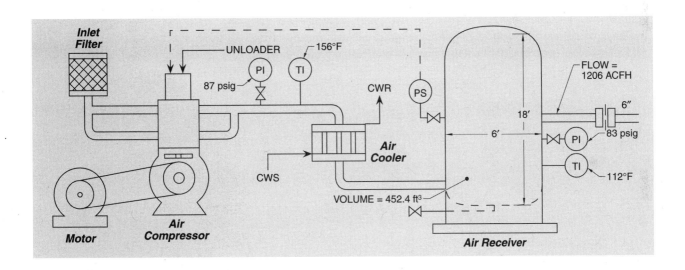

Calculate the time it takes for the pressure in the receiver to drop from 83 psig to 60 psig when the compressor shuts off. The outlet flow rate is 7430 scfh at 14.7 psia and 70°F.

_____ **1.** Use the gas laws to calculate the standard volume of gas at 83 psig.

_____ **2.** Use the gas laws to calculate the standard volume of gas at 60 psig.

_____ **3.** Calculate the change in standard volume of gas between the initial and final conditions.

_____ **4.** Calculate the time it takes for the pressure to drop from 83 psig to 60 psig.

SECTION
5 *FLOW MEASUREMENT*

Differential Pressure Flowmeters

chapter
19

REVIEW
QUESTIONS

Name _____ **Date** _____

T   F   **1.** A low-loss flow tube is a primary flow element consisting of an aerodynamic internal cross section with the low-pressure connection at the throat.

T   F   **2.** A blocking valve is a valve used at a differential measuring instrument to equalize high- and low-pressure sides of the differential instrument.

_____   **3.** A(n) ___ is a pipeline restriction that causes a pressure drop used to measure flow.
     A. venturi tube
     B. orifice plate
     C. flow nozzle
     D. all of the above

_____   **4.** A(n) ___ is a flow element consisting of a small bent tube with a nozzle opening facing into the flow.

T   F   **5.** The vena contracta is the point of lowest pressure and highest velocity downstream from a primary flow element.

_____   **6.** A(n) ___ is a primary flow element consisting of a thin circular metal plate with a sharp-edged round hole in it and a tab that protrudes from the flanges.
     A. flow nozzle
     B. orifice plate
     C. pitot tube
     D. venturi tube

_____   **7.** A(n) ___ pitot tube is a pitot tube consisting of a tube with several impact openings inserted through the wall of the pipe or duct and extending across the entire flow profile.

_____   **8.** The ___ equation states that the sum of the heads of an enclosed flowing fluid is the same at any two locations.

_____   **9.** A(n) ___ is a pressure connection.

T   F   **10.** Pressure-sensing taps are located in piping a fixed distance upstream and downstream of a flow nozzle.

_____   **11.** ___ is the ratio of maximum flow to minimum measurable flow at a desired measurement accuracy.
     A. Turndown
     B. Differential pressure
     C. Rangeability
     D. all of the above

_____ **12.** A(n) ___ is a primary flow element consisting of a restriction shaped like a curved funnel that allows a little more flow than other primary flow elements and reduces the straight run pipe requirements.

T   F   **13.** Flow measurement is only accurate as long as the flowing conditions remain the same as when the system was designed.

_____ **14.** A(n) ___ is a primary flow element consisting of a fabricated pipe section with a converging inlet section, a straight throat, and a diverging outlet section.

T   F   **15.** When liquid flow is measured, the presence of air bubbles in the impulse lines improves the measurement accuracy.

_____ **16.** A pressure difference is created when a fluid passes through a(n) ___ in a pipe.

_____ **17.** A(n) ___ is the tubing or piping connection that connects flowmeter taps to any differential pressure instruments.
    A. flange tap
    B. impulse line
    C. pipe tap
    D. vena contracta tap

_____ **18.** The ___ ratio is the ratio of maximum measurable value to minimum measurable value that can still produce full-scale output.

_____ **19.** Bernoulli determined that ___ was present at any point in a closed pipe.
    A. static head due to applied pressure
    B. static head due to elevation
    C. velocity head
    D. all of the above

_____ **20.** When steam flow is measured, the measuring instrument must be located ___ the flow element.
    A. above
    B. below
    C. at the same level as
    D. none of the above

Name _____ **Date** _____

Activity 19-1—Pressure Drops

An orifice plate is installed between a pair of flanges. The flow is from left to right. There are vertical reference lines provided at the 2½ D upstream point, the vena contracta point, and the 8 D downstream point.

Use the following drawing and chart to complete the tasks.

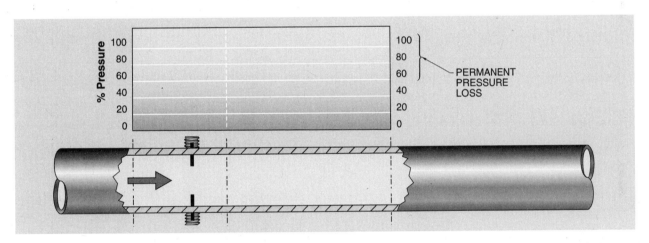

1. Draw the typical flow stream lines that pass through the pipe and orifice.

2. On the scale above the pipe line and orifice is a graduated chart. On this chart, draw the representative pressure value to match the flow stream. Use a starting value of 100% at the left of the chart.

Name _____   **Date** _____

Activity 19-2—Correction Factors

An air flow instrument has an indicated flow of 6400 scfh. The actual measured pressure is 100 psig and the temperature is 125°F. The design pressure is 83 psig and the temperature is 112°F.

_____   **1.** What is the pressure correction factor?

_____   **2.** What is the temperature correction factor?

_____   **3.** What is the correct flow?

Name _____   **Date** _____

Activity 19-3—Flow Equations

A 30″ WC U-tube manometer is installed across an orifice installed in an air line from an air receiver. A design flow rate of 7430 scfh of air flowing through the orifice develops a 25″ WC pressure differential. The flow is proportional to the square root of the pressure drop across the orifice.

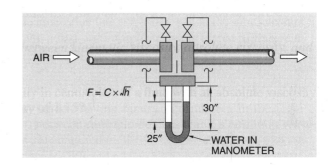

AIR

$F = C \times \sqrt{h}$

30″

25″ — WATER IN MANOMETER

Use a manometer to measure the pressure drop across the orifice. The pressure drop is equal to the height of the liquid column in the manometer.

_____   1. Calculate the value of C in the equation using the flow, F, in scfh and the pressure differential, h, in in. WC.

_____   2. What is the equation used to relate the pressure differential to the flow for this orifice/manometer combination?

3. Develop a chart showing the flow value for manometer readings of 0″, 5″, 10″,15″, 20″, 25″, 30″.

h	$\sqrt{h}$	F (scfh)
0	_____	_____
5	_____	_____
10	_____	_____
15	_____	_____
20	_____	_____
25	_____	_____
30	_____	_____

Replace the manometer with a differential pressure gauge with a linear scale readout of 0 to 100. The differential measurement instrument has a maximum differential of 30.0 in. WC.

_____ 4. Calculate the flow factor to be multiplied by the scale reading to obtain the flow rate.

_____ 5. What is the equation used to relate the scale reading to the flow?

Replace the linear pressure gauge with a square root scale readout of 0 to 10. The differential measurement instrument has a maximum differential of 30.0 in. WC.

_____ 6. Calculate the flow factor to be multiplied by the scale reading to obtain the flow rate.

_____ 7. What is the equation used to relate the scale reading to the flow?

SECTION
5 FLOW MEASUREMENT

chapter
20
Mechanical Flowmeters
REVIEW
QUESTIONS

Name _____   **Date** _____

_____ **1.** A(n) ___ rotameter consists of a tapered metal tube and a rod-guided float.

T   F   **2.** Manufacturers of rotameters provide capacity tables listing the float and tube for specified flow.

_____ **3.** A(n) ___ is a variable-area flowmeter consisting of a tapered tube and a float with a fixed diameter.
     A. orifice
     B. pitot tube
     C. rotameter
     D. none of the above

_____ **4.** A ___ is a small metal or plastic rotameter with an adjustable valve at the inlet or outlet of the meter to control the flow rate of the purge fluid.
     A. purge meter
     B. pitot tube
     C. Bernoulli meter
     D. bypass meter

T   F   **5.** A differential pressure flowmeter maintains a variable flow area and measures absolute pressure.

_____ **6.** ___ is a force that acts on the float of a rotameter.
     A. Pressure
     B. Gravity
     C. Rotation
     D. all of the above

_____ **7.** A(n) ___ rotameter consists of a clear plastic tube that allows visual observation of flow rate.

T   F   **8.** A rotameter can only provide correct flow rates for compressible gases and vapors when flowing conditions are the same as design conditions.

_____ **9.** A(n) ___ flowmeter maintains a constant differential pressure and allows the flow area to change with flow rate.

T   F   **10.** Rotameters consist of a float that moves in a tapered tube.

_____ **11.** A(n) ___ rotameter consists of a clear glass tube to allow visual observation of the flow rate.

_____ 12. A ___ meter is a flowmeter consisting of an orifice as part of the float assembly that acts as a guide.
   A. purge
   B. bypass
   C. shaped-float and orifice
   D. none of the above

_____ 13. A(n) ___ meter is a combination of a rotameter and an orifice plate used to measure flow rates through large pipes.

T   F   14. An orifice flowmeter maintains a constant differential pressure and allows area to change with flow rate.

_____ 15. ___ meters are common types of modified rotameters.
   A. Purge
   B. Bypass
   C. Shaped-float and orifice
   D. all of the above

_____ 16. A(n) ___ meter is a flowmeter consisting of a straight tube and a tapered cone with an indicator that moves up and down the cone with changes in flow.

_____ 17. A(n) ___ is a mechanical flowmeter consisting of turbine blades mounted on a wheel that measure the velocity of a liquid stream by counting the pulses produced by the blades as they pass an electromagnetic pickup.
   A. rotating disk meter
   B. rotameter
   C. turbine meter
   D. orifice meter

_____ 18. A(n) ___ meter is a positive-displacement flowmeter where fluid fills a chamber formed by sliding vanes mounted on a common hub rotated by the fluid.

_____ 19. A(n) ___ meter is a positive-displacement flowmeter for liquids where liquid flows through chambers, causing a disk to rotate and wobble.

_____ 20. A ___ is a special form of open-channel flow element that requires much less channel elevation change than a standard weir.
   A. Parshall weir
   B. Parshall flume
   C. Cipolleti weir
   D. Cipolleti flume

Name _____ Date _____

Activity 20-1—Rotameter Selection

The proper selection of a rotameter from a manufacturer's catalog can be more confusing and difficult than expected. The capacity values shown in manufacturer's catalogs are for water or for air flowing at 14.7 psia and 70°F.

If a liquid to be measured has a higher specific gravity than water, the additional buoyancy on the float results in reducing the effective weight of the float; at the same time, the higher specific gravity increases the momentum of the flowing liquid. The result is that it takes a lower flow to lift the float to the top of the scale when compared to water. When measuring gas flow, the gases are almost always at a higher pressure than 14.7 psia and are most likely to be at a different temperature. The gas may not even be air, which then changes the gas specific gravity. Gas at higher pressure means that more scfm of gas must pass through the rotameter to lift the float to the top of the scale.

In some cases it may be necessary to use a float made of material other than 316 SS. This changes the float specific gravity and weight.

Float Specific Gravity
316 SS = 8.02   Monel® = 8.84
Hastelloy® C = 8.94   Teflon® = 2.31

Liquid Flow. The change in liquid flow caused by a change in liquid specific gravity is calculated as follows:

$$F_2 = F_1 \times \sqrt{\frac{(\rho_f - \rho_2) \times \rho_1}{(\rho_f - \rho_1) \times \rho_2}}$$

where
- F_1 = original flow
- F_2 = new flow
- ρ_f = float specific gravity
- ρ_1 = original fluid specific gravity (water or air)
- ρ_2 = new fluid specific gravity

The change in liquid flow caused by a change in float specific gravity and/or weight is calculated as follows:

$$F_2 = F_1 \times \sqrt{\frac{W_{f2} \times (\rho_{f2} - \rho) \times \rho_{f1}}{W_{f1} \times (\rho_{f1} - \rho) \times \rho_{f2}}}$$

where
- F_1 = original flow
- F_2 = new flow
- W_{f1} = original float weight
- W_{f2} = new float weight
- ρ_1 = original float specific gravity
- ρ_2 = new float specific gravity
- ρ_f = fluid specific gravity

Gas Flow. The calculations for changes in gas flow need to take into consideration any changes in pressure, temperature, and specific gravity from the original conditions. If the gas composition changes, resulting in a change in the specific gravity, the new flow is scfm of the new gas, not the original gas. The "original" in the equations is the starting conditions. This may be the flow listed in the catalog or the actual process conditions. The "new" in the equations is the condition of flow to be determined. In either case, the flow is expressed in scfm. The corrections are calculated as follows:

$$\text{Flow(new)} = \text{Flow(original)} \times PC \times TC \times GC$$

where

pressure correction = $\quad PC = \sqrt{\dfrac{14.7 + psig(new)}{14.7 + psig(original)}}$

temperature correction = $\quad TC = \sqrt{\dfrac{460 + {}^\circ F(original)}{460 + {}^\circ F(new)}}$

specific gravity correction = $\quad GC = \sqrt{\dfrac{SG(original)}{SG(new)}}$

Nominal Capacity		Connections, IPS	Min. Operating Pressure for Gas Service, psig	Viscosity Ceiling, CSS	Pressure Drop, Inches Water	Tube Size, Inches	Tube Number	Float Number
gpm water	scfm air							
2.0	8.9	½	0	3.8	20	½	½-40-F-10	½-G-6YI-13
2.5	10.8	½	0	3.5	30	½	½-40-F-10	½-G-8YI-16
2.9	12.7	½	30	3.0	44	½	½-40-F-10	½-G-10YI-24
2.9	12.4	¾	0	3.0	5.0	¾	¾-40-F-10	¾-G-2YI-15
3.6	16.2	½	30	3.8	63	½	½-40-F-10	½-G-12YI-29
3.5	15.1	¾	0	3.1	7.6	¾	¾-40-F-10	¾-G-4YI-21
4.2	19.0	½	30	3.5	81	½	½-40-F-10	½-G-14YI-32
4.0	17.8	¾	0	3.2	10	¾	¾-40-F-10	¾-G-6YI-28

FLOAT WEIGHT NUMBER (316 SS)

NOTE: All values assume the use of stainless steel floats.

Use the catalog information to answer the following questions.

_____ **1.** What is the maximum flow of the 4.0 gpm rotameter shown in the sample catalog if it is used with a liquid with a specific gravity of 1.1?

_____ **2.** What is the maximum flow of the 3.5 gpm rotameter if it is to be used to measure brine flow with a specific gravity of 1.2? The 316 SS float must be replaced with a Monel float of the same size. Calculate the new float weight using the ratio of the float specific gravities.

_____ **3.** It is desired to measure 2.3 gpm of brine with a glass tube rotameter. The rotameter has a fluid specific gravity of 1.20. Select a 10″ rotameter with a 316 SS float so that the measured flow is about 75% of the rotameter maximum capacity.

_____ **4.** Select a 10″ glass tube rotameter from the catalog list that is able to measure a maximum flow of 37.5 scfm air at a pressure of 60.0 psig and 95°F.

Name _____   Date _____

Activity 20-2—Paddle Wheel Meter Selection

A 4″ pipeline handles a flow rate of 400 gpm of water. It is desired to use a paddle wheel flowmeter to have a flow range of up to 500 gpm. The meter needs to measure a minimum flow of 50 gpm.

Pipe Size (in.)	Std. (1)	Opt.	*Min.	Total
0.50	0-18	0-12	0-8	× 1
0.75	0-30	0-18	0-12	× 1
1	0-50	0-30	0-18	× 1
1.25	0-80	0-50	0-30	× 10
1.50	0-120	0-80	0-50	× 10
2	0-180	0-120	0-80	× 10
2.50	0-300	0-180	0-120	× 10
3	0-500	0-300	0-180	× 10
4	0-800	0-500	0-300	× 100
5	0-1200	0-800	0-500	× 100
6	0-1800	0-1200	0-800	× 100
8	0-3000	0-1800	0-1200	× 100
10	0-5000	0-3000	0-1800	× 100
12	0-8000	0-5000	0-3000	× 1000
14	0-12000	0-8000	0-5000	× 1000
16	0-12000	0-8000	0-5000	× 1000
18	0-18000	0-12000	0-8000	× 10000

NOTE 1: Flow rangeability is 20:1.

_____   **1.** Select the best paddle wheel meter size and range using the manufacturer's catalog information.

SECTION

5 **FLOW MEASUREMENT**

Magnetic, Ultrasonic, and
Mass Flowmeters

chapter

21

REVIEW
QUESTIONS

Name _____   **Date** _____

T   F   **1.** A mass flowmeter is an electronic flowmeter that uses the principle of sound transmission in liquids.

_____   **2.** A ___ meter is a flowmeter consisting of a stainless steel tube lined with nonconductive material, with two electrical coils mounted on the tube like a saddle.
   A. magnetic
   B. saddle
   C. turbine
   D. vortex shedding

T   F   **3.** A vortex shedding meter is a flowmeter consisting of a pipe section with a symmetrical vertical bluff body (a partial dam) across the flowing stream.

T   F   **4.** An ultrasonic flowmeter measures the actual mass of a flowing fluid.

_____   **5.** A(n) ___ meter is a mass flowmeter consisting of specially formed tubing that is oscillated at a right angle to the flowing mass of fluid.
   A. Coriolis
   B. ultrasonic
   C. magnetic
   D. thermal mass

T   F   **6.** A thermal mass meter is an ultrasonic flowmeter consisting of two sets of transmitting and receiving crystals and a cooling element that measures the heat loss to the fluid mass.

_____   **7.** A ___ ultrasonic meter is a meter consisting of two sets of transmitting and receiving crystals, one set aimed diagonally upstream and the other aimed diagonally downstream.
   A. diagonal
   B. Coriolis
   C. transit time
   D. Bernoulli

_____   **8.** With a vortex shedding meter, the frequency that the vortices are released from the bluff body can be measured with ___.
   A. thermal mass meters
   B. ultrasonic sensors
   C. differential pressure meters
   D. Coriolis meters

_____ **9.** The detectors for Coriolis meters consist of a(n) ___ mounted on each tubing section at the points of maximum motion.
    A. ultrasonic sensor
    B. turbulence meter
    C. magnet and coil
    D. vortex counter

_____ **10.** A Coriolis meter can also calculate ___ since the mass is measured and the volume is known.
    A. density
    B. viscosity
    C. temperature
    D. Doppler frequency

SECTION
5 FLOW MEASUREMENT

chapter
21

Magnetic, Ultrasonic, and
Mass Flowmeters

ACTIVITIES

Name _____   Date _____

Activity 21-1—Magnetic Flowmeter Selection

A wastewater stream has an operating flow range of 30 gpm up to 500 gpm through an 8″ pipe. To minimize the pressure loss through a magnetic flowmeter and any required inlet and outlet reducers, the largest meter that can measure the maximum and minimum flows should be selected.

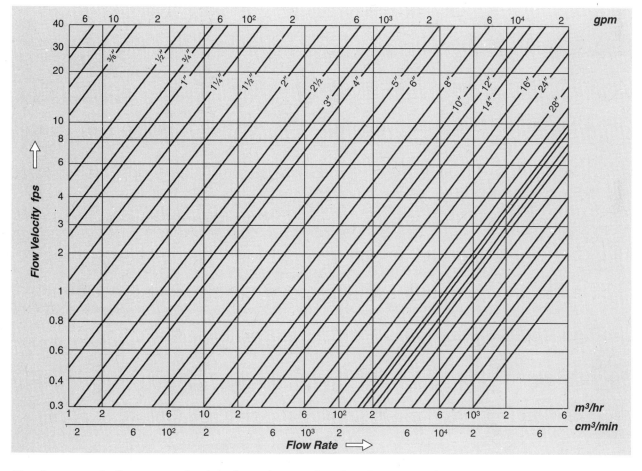

Use the magnetic flowmeter selection chart. Assume that the minimum and maximum measurable flow velocities are represented by the upper and lower limits of the chart.

_____ **1.** Select the largest meter that can measure down to a minimum flow rate of 30 gpm and up to a maximum flow rate of 500 gpm.

Name _____   **Date** _____

Activity 21-2—Vortex Shedding Meter Selection

A maximum flow of 1500 scfm at 100 psig and 95°F is to be measured. The minimum measured flow is 300 scfm. It was decided the meter should be able to read a range of 0 scfm to 2000 scfm. The line size is 4″ Schedule 40 pipe.

Nominal Capacity		Internal DIA	Minimum Air Flow	Maximum Air Flow			Maximum Water Flow	
inches	mm	inches	acfh	acfh	ft/sec	max Δp (psi)	gpm	max Δp (psi)
1.0	25	0.957	343	2895	161	0.70	79	16.25
1.5	40	1.500	742	11310	256	1.09	211	21.90
2.0	50	1.939	1519	15891	215	0.70	308	16.82
3.0	80	2.900	2649	35314	214	0.80	748	19.80
4.0	100	3.826	4238	67098	233	0.65	1188	16.40
6.0	150	5.761	8829	143023	219	0.90	2773	17.25
8.0	200	7.625	8829	282520	249	0.80	4843	17.10
10.0	250	9.562	28248	494340	275	0.80	7485	17.4
12.0	300	11.374	49434	706200	278	0.80	10568	17.7

_____   **1.** What is the acfh equivalent of the maximum flow rate to be measured?

_____   **2.** What is the acfh equivalent of the minimum flow rate to be measured?

_____   **3.** Select the best vortex shedding meter size using the sizing chart above.

_____   **4.** What is the correct flow in scfm if the indicated flow is 1650 scfm at a pressure of 85 psig?

SECTION
5 FLOW MEASUREMENT

chapter
22

Practical Flow Measurement

REVIEW
QUESTIONS

Name _____   **Date** _____

_____   **1.** A(n) ___ is a calculating device that totalizes the amount of flow during a specified time period.
   A. derivative
   B. integrator
   C. differential pressure switch
   D. mass flowmeter

_____   **2.** A(n) ___ switch is a flow switch consisting of a heated temperature sensor.
   A. orifice
   B. thermal
   C. thermocouple
   D. RTD

_____   **3.** A ___ switch is a flow switch consisting of a thin, flexible piece of metal inserted into a pipeline.
   A. blade
   B. thermal
   C. differential pressure
   D. rotameter

_____   **4.** A rotameter switch is a flow switch with a ___.
   A. shaped float
   B. fixed orifice
   C. magnet
   D. all of the above

_____   **5.** A(n) ___ switch is a flow switch consisting of a pair of pressure-sensing elements and an adjustable spring that can be set at a specific value to operate an output switch.
   A. differential pressure
   B. blade
   C. orifice
   D. thermal

_____   **6.** When gas flow is being measured, the measuring instrument must be mounted ___ the flow element.
   A. above
   B. so the impulse lines pass below
   C. below
   D. next to

_____ 7. A(n) ___ is the tubing or piping connection that connects the flowmeter taps to any of the differential pressure instruments.
   A. impulse line
   B. flow nozzle
   C. venturi
   D. orifice

_____ 8. When steam flow is being measured, the measuring instrument must be mounted ___ the flow element.
   A. beside
   B. at the height of
   C. above
   D. below

_____ 9. Differential pressure connections are typically made with flange, vena contracta, and ___ taps.
   A. shaped
   B. manifold
   C. pipe
   D. blocking

_____ 10. A(n) ___ valve is used at a differential pressure measuring instrument to provide a convenient location to isolate the instrument from the impulse, equalizing, or venting lines and to provide a way to equalize the high- and low-pressure sides of the differential pressure instrument.
   A. vapor
   B. steam
   C. pipe
   D. blocking

Name _____ **Date** _____

Compressor Capacity

An air compressor has an atmospheric inlet pressure. The air is compressed, cooled, and fed into an air receiver. Air from the receiver passes through an orifice plate flowmeter. The pressure in the receiver is 83 psig and the temperature is 112°F. The design flow at the location of the flowmeter is 1206 acfh at 83 psig and 112°F. The pressure switch controlling the compressor starts the compressor at a pressure of 83 psig and shuts the compressor off at 85 psig. The OFF cycle is 30 sec and the ON cycle is 2.0 minutes.

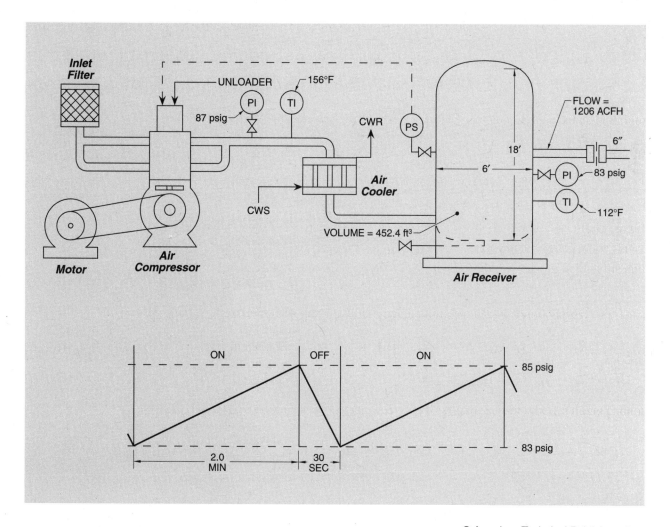

Calculate the compressor pumping capacity. The volume that is compressed in 2.0 minutes when the compressor is ON is equal to the amount that discharges during the same 2.0 minute period plus the amount that discharges during the 30 second period when the compressor is OFF. The design flow conditions are atmospheric pressure and 70°F.

_____ **1.** What is the flow rate, in scfh, out of the air receiver?

_____ **2.** What is the volume of flow, in scf, out of the receiver during the 2.5 minute cycle?

_____ **3.** What is the amount of flow, in scf, out of the receiver in 1 hour?

_____ **4.** What fraction of the time is the compressor operating?

_____ **5.** What is the compressor capacity?

SECTION
6 ANALYZERS

chapter
23

Gas Analyzers

REVIEW
QUESTIONS

Name _____   Date _____

_____ **1.** A(n) ___ analyzer is an analyzer consisting of a sensing head that uses magnetic fields to generate a "magnetic wind" that carries an oxygen-containing gas sample through a sample cell and across a pair of thermistors.

_____ **2.** A ___ analyzer is a gas analyzer that uses the principle that different gases absorb different, very specific, wavelengths of electromagnetic radiation in the infrared (IR) or ultraviolet (UV) regions of the electromagnetic spectrum.
    A. paramagnetic
    B. thermoparamagnetic
    C. radiant-energy absorption
    D. zirconium

_____ **3.** A common type of oxygen analyzer is a ___ analyzer.
    A. paramagnetic
    B. thermoparamagnetic
    C. zirconium oxide
    D. all of the above

_____ **4.** Analysis is the process of measuring ___ properties of chemical compounds so that the composition and quantities of the components can be determined.
    A. physical
    B. chemical
    C. electrical
    D. all of the above

_____ **5.** A(n) ___ analyzer is a gas analyzer that measures the concentration of a single gas in a sample by comparing its ability to conduct heat to that of a reference gas.
    A. infrared
    B. electrical conductivity
    C. thermal conductivity
    D. radiant-energy absorption

_____ **6.** A(n) ___ analyzer is an analyzer that measures an electric current generated by the reaction of oxygen with an electrolytic reagent.

  T   F   **7.** A nondispersive infrared (NDIR) analyzer is a radiant-energy absorption analyzer consisting of an IR electromagnetic radiation source, an IR detector, and two IR absorption chambers.

_____ **8.** A(n) ___ analyzer is an instrument that is located in the process area and obtains frequent or continuous samples from the process.

T   F   **9.** A zirconium oxide analyzer is a radiant-energy absorption analyzer consisting of a UV electromagnetic radiation source, a sample cell, and a detector that measures absorption of UV radiation by specific molecules.

T   F   **10.** An analyzer is an instrument used to provide an analysis of a sample from a process.

T   F   **11.** An opacity analyzer can be used to monitor for any particulate breakthroughs in a filtration or precipitation process.

T   F   **12.** All gases conduct heat.

_____ **13.** A(n) ___ analyzer is an analyzer consisting of two diamagnetic spheres filled with nitrogen that are connected with a bar to form a "dumbbell" shaped assembly.

T   F   **14.** Thermal conductivity analyzers work well with a mixture of three gases that vary in concentration.

_____ **15.** When paramagnetic force is ___, colder oxygen molecules displace heated oxygen molecules in the magnetic field.
        A. constant
        B. increased
        C. reduced
        D. none of the above

T   F   **16.** Thermal conductivity analyzers work best with gases that have similar thermal conductivities.

T   F   **17.** A paramagnetic oxygen analyzer measures electric current generated when the analyzer is subjected to different oxygen concentrations on opposite sides of an electrode.

_____ **18.** A(n) ___ analyzer is a gas analyzer consisting of a collimated (focused beam) light source and an analyzer to measure received light intensity.
        A. opacity
        B. electrochemical
        C. NDIR
        D. radiant-energy

_____ **19.** ___ analyzers are the most complex instruments in a plant.

_____ **20.** A zirconium oxide oxygen analyzer needs to operate at temperatures above ___°F.
        A. 50
        B. 100
        C. 500
        D. 1000

Name _____   Date _____

Activity 23-1—Thermal Conductivity

Thermal conductivity analyzers can be used to analyze the composition of binary gas samples. A binary gas is a gas mixture with only two gases in the sample. The thermal conductivity analysis method can only be used for binary gases with significantly different thermal conductivities or in pseudobinary gas mixtures where all but two of the gases have a fixed concentration. Thermal conductivity methods cannot be used to analyze the gas when more than two components vary in concentration. Each of the sections below concerns a gas mixture with one or more varying components given as volume percent. The gas component to be measured is the last gas in the list.

Gas	Relative Conductivity Air = 1
Air	1.00
Carbon dioxide (CO_2)	0.64
Helium (He)	5.79
Hydrogen (H_2)	7.17
Methane (CH_4)	1.32
Neon (Ne)	1.86
Nitrogen (N_2)	1.00
Oxygen (O_2)	1.02
Sulfur dioxide (SO_2)	0.37

Use the following gas mixture to answer questions 1–4.

Nitrogen (N_2) = 79%
Oxygen (O_2) = 5% to 20%
Carbon dioxide (CO_2) = 1% to 16%

_____   **1.** Is the above mixture suitable to be measured with a thermal conductivity gas analyzer?

2. State the reason for your answer.

_____

_____   **3.** If the mixture is suitable, calculate the relative thermal conductivity of the mixture at minimum and maximum concentrations.

_____   **4.** If the mixture is suitable, calculate the percent change in relative thermal conductivity from the low to high concentration.

Use the following gas mixture to answer questions 5–8.

Nitrogen (N_2) = 79% to 95%
Carbon dioxide (CO_2) = 1% to 19%
Oxygen (O_2) = 2% to 5%

_____   **5.** Is the above mixture suitable to be measured with a thermal conductivity gas analyzer?

6. State the reason for your answer.

_____

_____ **7.** If the mixture is suitable, calculate the relative thermal conductivity of the mixture at minimum and maximum concentrations.

_____ **8.** If the mixture is suitable, calculate the percent change in relative thermal conductivity from the low to high concentration.

Use the following gas mixture to answer questions 9–12.
   Air = 0% to 10%
   Hydrogen (H_2) = 90% to 100%

_____ **9.** Is the above mixture suitable to be measured with a thermal conductivity gas analyzer?

10. State the reason for your answer.

_____

_____ **11.** If the mixture is suitable, calculate the relative thermal conductivity of the mixture at minimum and maximum concentrations.

_____ **12.** If the mixture is suitable, calculate the percent change in relative thermal conductivity from the low to high concentration.

Use the following gas mixture to answer questions 13–16.
   Hydrogen (H_2) = 88% to 100%
   Nitrogen (N_2) = 0% to 10%
   Oxygen (O_2) = 0% to 2%
   Note: The N_2 and O_2 always vary in proportion to one another.

_____ **13.** Is the above mixture suitable to be measured with a thermal conductivity gas analyzer?

14. State the reason for your answer.

_____

_____ **15.** If the mixture is suitable, calculate the relative thermal conductivity of the mixture at minimum and maximum concentrations.

_____ **16.** If the mixture is suitable, calculate the percent change in relative thermal conductivity from the low to high concentration.

Use the following gas mixture to answer questions 17–20.
   Nitrogen (N_2) = 79%
   Air = 20% to 5%
   Carbon dioxide (CO_2) = 0% to 19%
   Natural gas (CH_4) = 0% to 2%

_____ **17.** Is the above mixture suitable to be measured with a thermal conductivity gas analyzer?

18. State the reason for your answer.

_____

_____ **19.** If the mixture is suitable, calculate the relative thermal conductivity of the mixture at minimum and maximum concentrations.

_____ **20.** If the mixture is suitable, calculate the percent change in relative thermal conductivity from the low to high concentration.

SECTION
6 ANALYZERS

chapter
24

Humidity and Solids
Moisture Analyzers

REVIEW
QUESTIONS

Name _____   Date _____

T   F   **1.** Wet bulb temperature is the highest temperature that can be obtained through the heating effect of water evaporating into the atmosphere.

_____   **2.** A ___ psychrometer consists of two glass thermometers attached to an assembly that permits the two thermometers to be rotated through the air.
A. sling
B. recording
C. hygrometer
D. double

_____   **3.** ___ is a ratio of the actual amount of water vapor in air to the maximum amount of water vapor possible at the same temperature.

_____   **4.** A(n) ___ analyzer is an instrument that measures the amount of moisture in air.

T   F   **5.** A thermohygrometer is a combination of a hygrometer, pressure sensor, and temperature-sensing instrument.

T   F   **6.** Absolute humidity is another name for humidity ratio.

_____   **7.** A ___ is a humidity analyzer that measures physical or electrical changes that occur in various materials as they absorb or release moisture.
A. sling psychrometer
B. hygrometer
C. recording psychrometer
D. dew cell

_____   **8.** ___ humidity is a ratio of mass of water vapor to the mass of dry air plus moisture.
A. Absolute
B. Relative
C. Specific
D. none of the above

_____   **9.** ___ is the amount of water vapor in a given volume of air.
A. Density
B. Humidity
C. Buoyancy
D. Saturation

_____ **10.** ___ are the band of electromagnetic radiation between infrared and VHF broadcast frequencies, covering the range of approximately 3 mm to 3 m wavelengths.

T   F   **11.** Radiowave moisture analyzers are available as either reflection type or transmission type.

_____ **12.** The continuous measurement of moisture content in solids is crucial to the mass production of granular ___.
   A. chemicals
   B. grains
   C. plastics
   D. all of the above

T   F   **13.** A noncontact detection method is required for determining moisture content of granular materials.

_____ **14.** ___ is the ability of a material to reflect light or radiant energy.

_____ **15.** The two methods most suited to continuous high-speed processing are based on the use of ___ radiation and ___.
   A. near infrared; microwaves
   B. near infrared; radiowaves
   C. far infrared; microwaves
   D. far infrared; radiowaves

_____ **16.** A(n) ___ moisture analyzer is an analyzer that measures the amount of moisture in solids by measuring the speed of neutrons that strike the object.
   A. electrical impedance
   B. nuclear solids
   C. microwave
   D. radiowave

_____ **17.** A near infrared moisture analyzer is a solids moisture analyzer that measures the ___ of the process material and calculates moisture content.
   A. transmittance
   B. temperature
   C. reflectance
   D. admittance

Name _____   **Date** _____

Activity 24-1—Psychrometric Charts

Use the psychrometric chart in the Appendix to answer the following questions.

Given a dry bulb of 72°F and a wet bulb of 65°F, find the corresponding values of the following:

_____   **1.** Relative humidity %

_____   **2.** Dewpoint °F

_____   **3.** Humidity ratio

Given a dewpoint of 78°F and a relative humidity of 81%, find the corresponding values of the following:

_____   **4.** Wet bulb °F

_____   **5.** Dry bulb °F

_____   **6.** Humidity ratio

Given a wet bulb of 75°F and a humidity ratio of 0.010, find the corresponding values of the following:

_____   **7.** Relative humidity %

_____   **8.** Dry bulb °F

_____   **9.** Dewpoint °F

Given a dry bulb of 45°F and a relative humidity of 65%, find the corresponding values of the following:

_____   **10.** Wet bulb °F

_____   **11.** Dewpoint °F

_____   **12.** Humidity ratio

Given a dry bulb of 95°F and a wet bulb of 90°F, find the corresponding values of the following:

_____   **13.** Relative humidity %

_____   **14.** Dewpoint °F

_____   **15.** Humidity ratio

SECTION
6 ANALYZERS

chapter
24

Humidity and Solids
Moisture Analyzers

ACTIVITIES

Name _____   Date _____

Activity 24-2—Humidification

The outside air has a dry bulb temperature of 47°F and a wet bulb temperature of 44°F. The flow rate of the air into the building is 117 scfm. One SCF of dry air weighs 0.0756 lb/cu ft. Use the psychrometric chart in the Appendix as needed to answer the following questions.

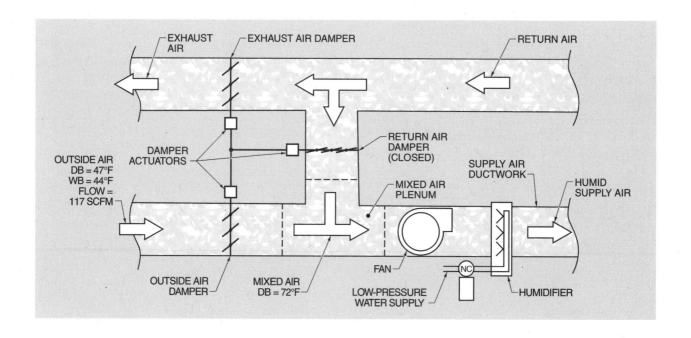

_____ 1. Determine the relative humidity of the outside air.

_____ 2. Determine the humidity ratio of the outside air.

_____ 3. This air is drawn into a building and then heated to 72°F. Determine the relative humidity at this temperature.

_____ 4. How much water needs to be added to each pound of the heated air to bring the relative humidity up to 50%?

_____ 5. How much water in lb/min needs to be added to the heated air to obtain 50% RH?

126

Name _____   Date _____

Activity 24-3—Dehumidification

One way that warm outside air can be dehumidified is by cooling the air. Passive dehumidification is the process of removing moisture from air by using the existing cooling coils of an HVAC system. Moisture condenses out of the air as the air is cooled.

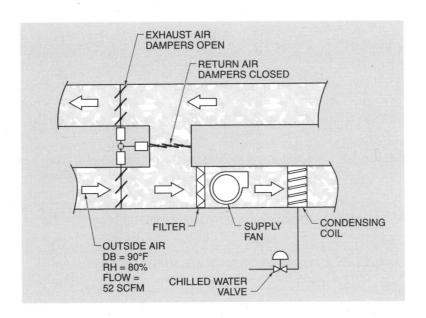

The outside air temperature is 90°F and the relative humidity is 80%. Use the psychrometric chart in the Appendix as needed to answer the following questions.

_____   **1.** Determine the humidity ratio for this outside air.

_____   **2.** The desired indoor air temperature is 72°F with 50% RH. To what temperature does the outside air have to be cooled to condense sufficient water out of the air to reach these conditions when reheated?

_____ **3.** What is the humidity ratio at a temperature of 72°F and 50% RH?

_____ **4.** How much water, in lb/min, needs to be removed from air flowing at 52 scfm to reach the desired conditions of 72°F and 50% RH?

SECTION
6 ANALYZERS

chapter
24
ACTIVITIES

Humidity and Solids
Moisture Analyzers

Name _____   Date _____

Activity 24-4—Calibration

A rotary dryer is used to dry a granular material. To reduce handling problems associated with static electricity, a small amount of moisture is left in the material. The material moisture content is measured with a noncontact moisture analyzer. It is necessary to check the instrument reading on a regular basis to ensure accuracy. A laboratory scale is used to check the reading.

A sample of the material is collected in a metal pan that can be sealed with a lid. The sample is taken to the lab where it is weighed including the lid. The lid is then removed and the pan and sample are placed in a drying oven and kept there long enough to ensure that the sample is completely dry. The sample is removed and weighed along with the lid. The material is dumped out and the empty pan and lid are measured again. The percent (wt %) moisture is calculated.

The reading of the solids moisture instrument is 4.2 wt %.
The weight of wet material, pan, and lid = 2015 g.
The weight of dry material, pan, and lid = 1982 g.
The weight of empty pan and lid = 1237 g.

_____ 1. What is the wt % of moisture in the original sample?

_____ 2. What is the measurement error percent relative to the actual value determined in the lab?

Another sample is taken from a different process.
The reading of the solids moisture instrument is 54.7 wt %.
The weight of wet material, pan, and lid = 2243 g.
The weight of dry material, pan, and lid = 1899 g.
The weight of empty pan and lid = 1036 g.

_____ 3. What is the wt % of moisture in the original sample?

_____ 4. What is the measurement error percent, relative to the actual value determined in the lab?

_____ 5. The same sample was placed in the oven again and dried more. The new weight of dry material, pan, and lid was 1827 g. Based on this measurement, what is the wt % of moisture in the original sample?

6. What are possible reasons for a difference between the first and the second drying?

_____

_____

_____

SECTION
6 ANALYZERS

chapter
25

Liquid Analyzers

REVIEW
QUESTIONS

Name _____ Date _____

_____ 1. The formula used to convert process fluid head pressure to the equivalent head of water is ___.
    A. $P = H \div SG$
    B. $P = SG \div H$
    C. $P = H \times SG$
    D. $P = H + SG$

_____ 2. A ___ hydrometer consists of a fixed-volume and fixed-weight cylinder heavier than fluid and supported by a spring or a lever connected to a torque tube that acts as a spring.
    A. continuous
    B. liquid density
    C. U-tube
    D. none of the above

_____ 3. A(n) ___ is a viscosity analyzer consisting of a heated, constant-temperature cylinder where a polymer is melted and forced through an orifice by a piston moving at a constant rate.

_____ 4. A liquid analyzer is an instrument that can be used to measure the ___ of a liquid.
    A. density
    B. turbidity
    C. viscosity
    D. all of the above

T   F 5. A vibrating U-tube density analyzer consists of a suitable radioactive isotope source producing gamma rays directed through a chamber containing the liquid to be measured.

_____ 6. ___ of a liquid is the amount of bending of a light beam as it moves between fluids.

T   F 7. A hydrometer is a liquid density analyzer with a sealed float consisting of a hollow, tubular glass cylinder, a scale, and weights.

_____ 8. A(n) ___ analyzer consists of a precision tube and a piston with a timed fall through a constant-temperature liquid.

T   F 9. A nuclear radiation density analyzer consists of a U-tube that is fixed at the open ends and filled with a liquid.

_____ 10. ___ is the volume of solids suspended in a slurry divided by the total volume of the slurry.

_____ **11.** The ___ angle of a liquid is the one angle at which there is no light refraction and all the light is reflected.
    A. absolute
    B. critical
    C. refraction
    D. none of the above

T   F   **12.** A continuous hydrometer consists of a float that is completely submerged in a liquid and has chains attached to the bottom to change the effective float weight to match changes in density of the liquid.

_____ **13.** ___ is resistance to flow.
    A. Density
    B. Pressure
    C. Turbidity
    D. Viscosity

_____ **14.** A(n) ___ viscometer is a laboratory apparatus consisting of a temperature-controlled vessel with an orifice in the bottom for measuring the kinematic viscosity of oils and other viscous liquids.

_____ **15.** A(n) ___ analyzer is used to measure the density of a slurry.

_____ **16.** A(n) ___ analyzer consists of an instrument where a measured quantity of a process sample is mixed with precise quantities of reagents and then quantitatively measured with a pH meter or by colorimetric or other type of detector.

_____ **17.** A(n) ___ is a liquid whose viscosity does not change with applied force.

_____ **18.** A(n) ___ analyzer measures the amount of suspended solids in a liquid by measuring the amount of light scattering from the suspended particles.

_____ **19.** A rotating spindle viscosity analyzer consists of a rotating spindle in a container of ___ at a controlled temperature.

_____ **20.** ___ is the science of the deformation and flow of matter.
    A. Psychrometry
    B. Hygrometry
    C. Turbidity
    D. Rheology

_____ **21.** A(n) ___ analyzer is a liquid analyzer that measures a liquid's resistance to flow under specific conditions.

T   F   **22.** Pseudo density is the density of a slurry, determined by the total weight of the slurry including solids, divided by the volume of slurry.

Name _____ Date _____

Activity 25-1—Specific Gravity Corrections

Specific gravity instruments often need to operate at process temperatures other than the standard conditions. Therefore, the measured specific gravity needs to be corrected to what it would be at standard conditions. Use the specific gravity temperature correction table to determine the correction factors and the correct specific gravity value.

Temperature, °F	Correction
20	0.0002
30	−0.0007
35	−0.0009
40	−0.0009
50	−0.0006
60	0.0000
70	0.0010
80	0.0023
90	0.0040
100	0.0060
110	0.0082
120	0.0105
130	0.0130
140	0.0158
150	0.0188
160	0.0220

For a process liquid with a specific gravity of 1.085 at a temperature of 100°F, determine the following:

_____ 1. What is the specific gravity correction?

_____ 2. What is the corrected specific gravity?

For a process liquid with a specific gravity of 1.119 at a temperature of 150°F, determine the following:

_____ 3. What is the specific gravity correction?

_____ 4. What is the corrected specific gravity?

For a process liquid with a specific gravity of 1.103 at a temperature of 30°F, determine the following:

_____ 5. What is the specific gravity correction?

_____ 6. What is the corrected specific gravity?

For a process liquid with a specific gravity of 1.025 at a temperature of 85°F, determine the following:

_____ 7. What is the specific gravity correction?

_____ 8. What is the corrected specific gravity?

For a process liquid with a specific gravity of 1.056 at a temperature of 135°F, determine the following:

_____ 9. What is the specific gravity correction?

_____ 10. What is the corrected specific gravity?

Name _____   Date _____

Activity 25-2—Specific Gravity Measurement

It is desired to measure the specific gravity of a process fluid in a vessel using a differential pressure transmitter. The process fluid may have any specific gravity in the range of 1.050 to 1.150. The upper and lower d/p cell connections are below the overflow level in the vessel. The distance between the two connections is 96″. The process fluid is corrosive to the d/p cell, but it is compatible with water, so water can be used to purge the d/p cell and the connecting tubes. The d/p cell is located 10″ below the lower connection.

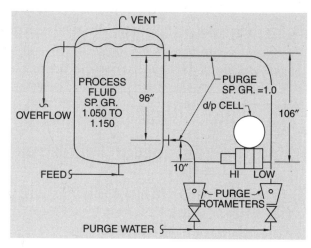

Assume that the vessel contains fluid at a specific gravity of 1.150, the upper range limit.

_____ **1.** Calculate the pressure head in inches WC on the d/p cell high-pressure connection.

_____ **2.** Calculate the pressure head in inches WC on the d/p cell low-pressure connection.

_____ **3.** Calculate the d/p cell differential.

Assume that the vessel contains fluid at a specific gravity of 1.050, the lower range limit.

_____ **4.** Calculate the pressure head in inches WC on the d/p cell high-pressure connection.

_____ **5.** Calculate the pressure head in inches WC on the d/p cell low-pressure connection.

_____ **6.** Calculate the d/p cell differential.

Perform the following calculations for the overall d/p system:

_____ **7.** Calculate the d/p cell calibration range and suppression or elevation.

8. What changes would occur to the d/p cell calibration if the transmitter were located at a 20″ lower elevation?

Name _____   Date _____

Activity 25-3—Viscosity Conversions

Calculate the following values based on the relationship between kinematic viscosity, absolute viscosity, and specific gravity.

_____   1. What is the kinematic viscosity in centistokes of a fluid with an absolute viscosity of 23 cP and a specific gravity of 1.15?

_____   2. What is the kinematic viscosity in centistokes of a fluid with an absolute viscosity of 1.0 cP and a specific gravity of 1.0?

_____   3. What is the kinematic viscosity in centistokes of a fluid with an absolute viscosity of 3015 cP and specific gravity of 0.876?

_____   4. What is the absolute viscosity in centipoise of a fluid with a kinematic viscosity of 35.3 cSt and a specific gravity of 0.85?

_____   5. What is the absolute viscosity in centipoise of a fluid with a kinematic viscosity of 1.15 cSt and a specific gravity of 0.78?

_____   6. What is the absolute viscosity in centipoise of a fluid with a kinematic viscosity of 2365 cP and a specific gravity of 0.952?

_____   **7.** What is the specific gravity of a fluid with an absolute viscosity of 50 cP and a kinematic viscosity of 54.3 cSt?

_____   **8.** What is the specific gravity of a fluid with an absolute viscosity of 2000 cP and a kinematic viscosity of 2160 cSt?

_____   **9.** What is the specific gravity of a fluid with an absolute viscosity of 3.15 cP and a kinematic viscosity of 3.43 cSt?

SECTION
6 ANALYZERS

chapter
26

Electrochemical and
Composition Analyzers

REVIEW
QUESTIONS

Name _____   **Date** _____

_____ 1. A representative sample is a sample from a process in which the composition of the sample is ___ the sample in the process piping.
   A. larger than
   B. smaller than
   C. the same as
   D. none of the above

_____ 2. A(n) ___ is the measurement of the acidity or alkalinity of a solution caused by the dissociation of chemical compounds in water.

_____ 3. A(n) ___ analyzer is an electrochemical analyzer that measures the electrical conductivity of liquids and consists of two electrodes immersed in a solution.

_____ 4. ___ is a type of liquid chromatography that uses high pressure to force the liquid sample through a column at a faster rate than the liquid would normally travel.
   A. Accelerated chromatography
   B. High-pressure liquid chromatography (HPLC)
   C. Composition chromatography
   D. Gas-phase chromatography

_____ 5. A ___ is the modern unit of electrical conductivity and is the reciprocal of resistance.
   A. lumen
   B. watt
   C. siemens
   D. joule

_____ 6. pH is usually measured on a scale of 0 to 14 with 7 being neutral, less than 7 being ___, and greater than 7 being ___.
   A. benzoic; acidic
   B. acidic; benzoic
   C. alkaline; acidic
   D. acidic; alkaline

_____ 7. A conductivity cell constant is a ratio of the size of actual ___ to those of a standard conductivity cell.

_____ 8. ORP stands for ___.
   A. oxidated reactant processing
   B. oxidation reduction potential
   C. oil rig propulsion
   D. oxygenated radical percentage

T   F   **9.** An unbuffered solution is a solution of a strong acid or strong base without any other chemicals that react with the acid or base.

_____ **10.** A(n) ___ analyzer is an analyzer used to measure the quantity of multiple components in a single sample from a process.
    A. composition
    B. dual
    C. ORP
    D. pH

_____ **11.** A(n) ___ is a graph that shows the quantities of reagent required to change the pH of a solution.

T   F   **12.** Analyzer calibration is a process of substituting known sample compositions for a normal process sample so that the analyzer can be adjusted to read the correct values.

_____ **13.** A(n) ___ analyzer is an electrochemical analyzer consisting of a metal measuring electrode and a standard reference electrode that measure voltage produced by an electrochemical reaction between metals of the electrodes and chemicals in solution.

T   F   **14.** An analyzer sampling system is the time that it takes for a sample to get from the process through the final analysis.

_____ **15.** A(n) ___ analyzer is an instrument consisting of a small stainless steel tube packed with a porous inert material such as silica gel or alumina, an injection valve assembly, and a detector.

T   F   **16.** A chromatographic column is a stainless steel tube or length of capillary tubing of a chromatography instrument after the tube is filled with packing and ready to use.

T   F   **17.** A strong acid, HCl, mixed with a weak base, Na_2CO_3, is an unbuffered solution.

_____ **18.** A(n) ___ analyzer is an electrochemical analyzer consisting of a cell that generates an electric potential when immersed in a sample.
    A. pH
    B. electrolytic
    C. cell
    D. chemical

T   F   **19.** A sample transportation lag is a system of piping, valves, and other equipment that is used to extract a sample from a process stream, condition it if necessary, and convey it to an analyzer.

_____ **20.** ___ is the process of separating components of a sample transported by an inert carrier stream through a variety of media.
    A. ORP
    B. Titration
    C. Chromatography
    D. Decomposition

SECTION
6 ANALYZERS

chapter
26

Electrochemical and
Composition Analyzers

ACTIVITIES

Name _____   Date _____

Activity 26-1—Conductivity

Conductivity is often used to determine the concentration of a solution. The following data is for ammonia, NH_3.

Conductivity, μS/cm	Concentration, wt %
6.6	0.0001
14	0.0003
27	0.001
49	0.003
84	0.01
150	0.03
275	0.1
465	0.3
810	1
1110	3
1120	10
210	30

Use the table above and graph paper on the back to answer the following questions.

1. On the back of this page, graph the conductivity of NH_3 from a concentration of 0.1 wt % through 30 wt %. Place conductivity on the x-axis and concentration on the y-axis.

2. What is the special property, or measurement difficulty, that this conductivity illustrates?

   _____

3. What kind of a problem would an instrument or transmitter have in measuring this concentration range?

   _____

4. What range of concentration would it be possible to measure, considering the special properties of the concentration curve?

   _____

5. Is this special property an isolated case, or is it more common?

   _____

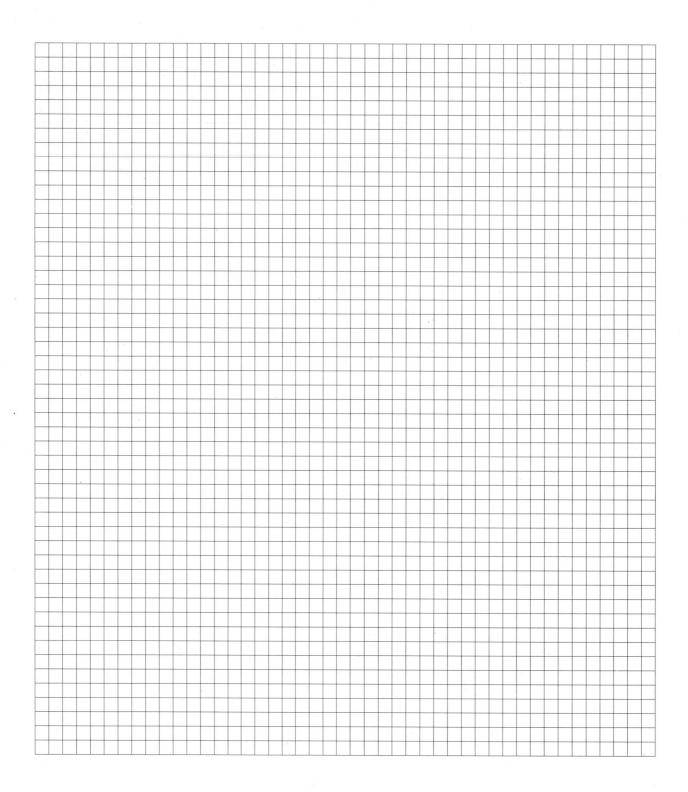

Electrochemical and Composition Analyzers

Name _____   Date _____

Activity 26-2—Concentration Limits

Conductivity is often used to determine the concentration of a solution. For most chemical solutions, the conductivity goes through a maximum as the concentration increases. The maximum measurable concentration normally occurs at the maximum conductivity.

Conductivity at 77°F*

% by Weight	PPM	NaCl	NaOH	H_2SO_4	Acetic Acid	CO_2	NH_3	H_3PO_4	SO_2
0.0001	1	2.2	6.2	8.8	4.2	1.2	6.6	—	—
0.0003	3	6.5	18.4	26.1	7.4	1.9	14	—	—
0.001	10	21.4	61.1	85.6	15.5	3.9	27	—	—
0.003	30	64	182	251	30.6	6.8	49	—	—
0.01	100	210	603	805	63	12	84	342	—
0.03	300	617	1780	2180	114	20	150	890	—
0.1	1000	1990	5820	6350	209	39	275	2250	3600
0.3	3000	5690	.0169	.0158	368	55	465	4820	7900
1	10,000	.0176	.0532	.0485	640	—	810	.0105	.0172
3	30,000	.0486	.144	.141	1120	—	1110	.0230	.0327
10	100,000	.140	.358	.427	1730	—	1120	.0607	.0610
30	300,000	—	.292	.822	1620	—	210	.182	—

*in microsiemens/cm
Shaded cells indicate values measured in siemens/cm.
Underscored values indicate conductivity passes through a maximum between the two listed concentrations.

Use the conductivity and concentration information shown in the data table to answer the following questions. Include the conductivity units in your answer.

_____   **1.** What is the maximum concentration of NaCl in wt % that can be measured before the conductivity begins decreasing?

_____   **2.** What is the conductivity of the NaCl solution at the maximum concentration?

_____   **3.** What is the maximum concentration of NaOH in percent by weight that can be measured before the conductivity begins decreasing?

_____   **4.** What is the conductivity of an NaOH solution at the maximum concentration?

_____ **5.** What is the maximum concentration of H_2SO_4 in percent by weight that can be measured before the conductivity begins decreasing?

_____ **6.** What is the conductivity of an NaOH solution at the maximum concentration?

_____ **7.** What is the maximum concentration of acetic acid in percent by weight that can be measured before the conductivity begins decreasing?

_____ **8.** What is the conductivity of an acetic acid solution at the maximum concentration?

_____ **9.** What is the maximum concentration of CO_2 in percent by weight that can be measured before the conductivity begins decreasing?

_____ **10.** What is the conductivity of a CO_2 solution at the maximum concentration?

_____ **11.** What is the maximum concentration of NH_3 in percent by weight that can be measured before the conductivity begins decreasing?

_____ **12.** What is the conductivity of an NH_3 solution at the maximum concentration?

_____ **13.** What is the maximum concentration of H_3PO_4 in percent by weight that can be measured before the conductivity begins decreasing?

_____ **14.** What is the conductivity of an H_3PO_4 solution at the maximum concentration?

_____ **15.** What is the maximum concentration of SO_2 in percent by weight that can be measured before the conductivity begins decreasing?

_____ **16.** What is the conductivity of an SO_2 solution at the maximum concentration?

17. Is it possible to measure higher solution concentrations than indicated by the conductivity maximum? If so, how can this be done?

_____

_____

SECTION

6 ANALYZERS

chapter

26

Electrochemical and
Composition Analyzers

ACTIVITIES

Name _____   Date _____

Activity 26-3—Conductivity Cell Constants

A conductivity cell constant is the ratio of the size of the actual electrodes to those of the standard conductivity cell. These constants range from 0.01 to 100. The larger cell constants are used to measure the more conductive solutions. Typical cell constants available are 0.01, 0.1, 1, 10, and 50. The standard conductivity cell is typically used over a conductivity range of 10 µS/cm to 2000 µS/cm. The conductivity ranges for the other cells are proportional to the change in the cell constant. For example, the conductivity range for a conductivity cell constant of 10 is 100 µS/cm to 20,000 µS/cm.

Calculate the conductivity range for the following cell constants.

_____   **1.** What is the conductivity range for a conductivity cell constant of 0.01?

_____   **2.** What is the conductivity range for a conductivity cell constant of 0.1?

_____   **3.** What is the conductivity range for a conductivity cell constant of 50?

A list of chemical solutions and conductivity ranges is shown below. List the correct cell constant that should be used with each of the measurements.

_____   **4.** Boiler feedwater, conductivity of 0.1 µS/cm to 2.0 µS/cm

_____   **5.** Return condensate, conductivity of 0.1 µS/cm to 5.0 µS/cm

_____   **6.** NaOH (0.1 to 10 wt %), conductivity of 5820 µS/cm to 358,000 µS/cm

_____   **7.** NaCl (0.01 to 1 wt %), conductivity of 210 µS/cm to 17,600 µS/cm

_____   **8.** CO_2 (0.03 to 0.3 wt %), conductivity of 20 µS/cm to 55 µS/cm

SECTION
6 ANALYZERS

chapter
26

Electrochemical and
Composition Analyzers

ACTIVITIES

Name _____ Date _____

Activity 26-4—Acid/Base Neutralization

Acid solutions have an excess of positively charged H^+ ions. Basic, or alkaline, solutions have an excess of negatively charged OH^- ions. The scale shown in the figure shows the relationships of H^+ ions and OH^- hydroxyl ions on opposite sides of the neutral point at a pH of 7.0. The number of H^+ ions on the acid side can be neutralized by an equal number of OH^- ions.

Use the relationships shown in the figure to answer the following questions.

1. How much 4% caustic soda is needed to neutralize 10 ml of 5% sulfuric acid?

2. How much vinegar, 4% acetic acid, is needed to neutralize 10 ml of 4% caustic soda?

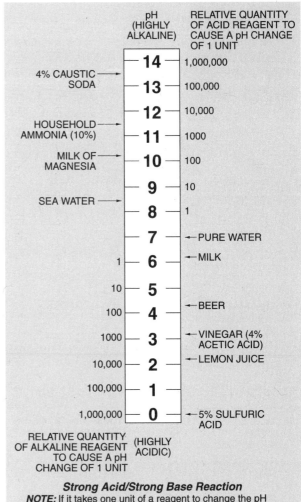

Strong Acid/Strong Base Reaction
NOTE: If it takes one unit of a reagent to change the pH of a solution from 7 to the next pH division, it takes 10 times as much reagent to change the pH to the next pH division.

SECTION
6 ANALYZERS

chapter
26

Electrochemical and
Composition Analyzers

ACTIVITIES

Name _____   Date _____

Activity 26-5—Wastewater Treatment

Wastewater treatment is a common industrial process. Water must be treated to neutralize the pH before the water is released from the treatment system. The figure shows a common strong acid/strong base titration curve. The titration curve was constructed by titrating a 500 ml wastewater sample with 1 normal caustic soda (1 N NaOH). With the wastewater starting at a pH of 2.0, it requires 28 ml of caustic soda solution to change the sample pH to 11.9.

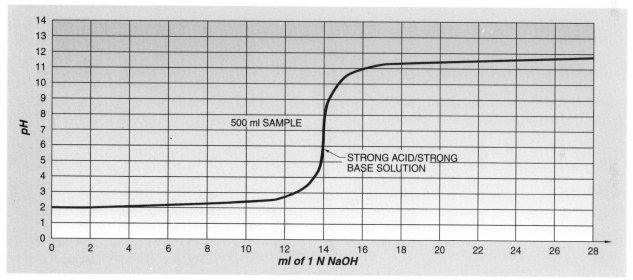

Use a 500 ml waste sample with an initial pH of 2.5 to answer the following questions.

_____ 1. How many ml of the 1 N NaOH are necessary to change the sample pH from the original 2.5 to 6.0?

_____ 2. How much more caustic soda would be needed to change the sample pH from 6.0 to 7.0?

_____ 3. What is the percentage of the total caustic soda required to change the sample pH from 6.0 to 7.0 compared to the total amount of caustic soda required to change the sample pH from 2.5 to 7.0?

4. Describe the effectiveness of a control valve at controlling the wastewater pH at 7.0.

_____

_____ 5. If the wastewater flow is 35 gpm, what flow of 1 N NaOH would be needed to change the wastewater from 2.5 pH to 7.0 pH?

SECTION
6 ANALYZERS

Electrochemical and
Composition Analyzers

chapter
26
ACTIVITIES

Name _____   Date _____

Activity 26-6—Buffered pH Neutralization

A waste stream containing carbonate chemicals changes the titration curve to a buffered titration curve. The figure shows a common buffered acid/base titration curve. The titration curve was constructed by titrating a 500 ml wastewater sample with 1 normal caustic soda (1 N NaOH). With the wastewater starting at a pH of 2.0, it requires 42 ml of caustic soda solution to change the sample pH to 10.9.

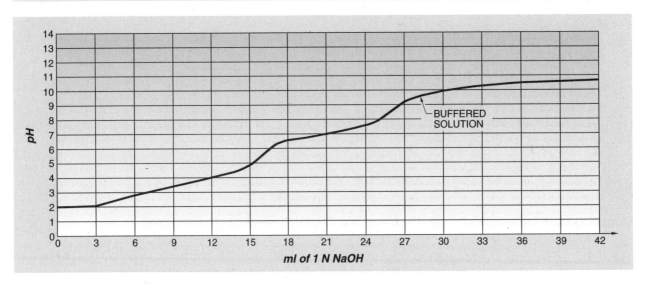

Use a 500 ml waste sample with an initial pH of 2.0 to answer the following questions.

_____   **1.** How much 1 N NaOH is needed to change the pH from 2.0 pH to 6.5 pH?

_____   **2.** How much 1 N NaOH is needed to change the pH from 6.5 pH to 7.0 pH?

_____   **3.** What percentage of caustic soda is needed to go from 6.5 pH to 7.0 pH as compared to the total caustic soda to go from 2.0 pH to 7.0 pH?

4. Describe the effectiveness of a control valve at controlling the wastewater pH at 7.0.

_____

5. How would the difficulty of control change if the control point were at a pH of 5.5 instead of a pH of 7.0?

_____

Name _____ **Date** _____

Differential Pressure

Salt from a centrifuge is added to a vessel along with saturated brine at a specific gravity of 1.2. The mixture is agitated to keep the salt in suspension. The pseudo specific gravity control point is 1.275 to ensure that the slurry is pumpable. The pseudo specific gravity measurement range is 1.2 to 1.3. The pseudo specific gravity is to be measured with a d/p cell with a double-filled capillary system and extended diaphragm seals. The seals are needed to prevent the vessel nozzle from filling with salt. The capillary filling fluid has a specific gravity of 1.5. The vessel specific gravity measurement connections are 120″ apart. The d/p cell transmitter is measured to 20″ below the lower measurement connection. A level controller (not shown) maintains the slurry level above the specific gravity measurement connections.

1. On the back of this page, make a drawing of the vessel, specific gravity measurement connections, d/p cell location, capillary lines, agitator, feeds, and all dimensions and specific gravities of process and capillary fluids.

For each of the following, assume that the specific gravity is at the upper range limit.

_____ 2. What is the pressure at the d/p cell high-pressure connection?

_____ 3. What is the pressure at the d/p cell low-pressure connection?

_____ 4. What is the d/p cell differential pressure?

For each of the following, assume that the specific gravity is at the lower range limit.

_____ 5. What is the pressure at the d/p cell high-pressure connection?

_____ 6. What is the pressure at the d/p cell low-pressure connection?

_____ 7. What is the d/p cell differential pressure?

Use the d/p cell differentials calculated for the upper and lower range limits for the following questions.

_____ **8.** What is the d/p cell range and suppression or elevation?

9. What changes would occur in the d/p cell calibration if the transmitter elevation were 20″ lower?

_____

_____

chapter
27

Mechanical and
Proximity Switches

REVIEW
QUESTIONS

Name _____   Date _____

_____ **1.** ___ switches require physical contact with an object to actuate the switch mechanism.
A. Mechanical
B. Ultrasonic
C. Photoelectric
D. Capacitance

T   F   **2.** Mechanical switches cannot be used in explosionproof areas.

_____ **3.** A(n) ___ proximity sensor consists of a sensor coil, an oscillator, a trigger circuit, and an output switching circuit.
A. mechanical
B. infrared
C. inductance
D. photoelectric

_____ **4.** A capacitance proximity sensor acts as a simple capacitor with one plate built into the sensor and the ___ being the other plate.

_____ **5.** Proximity sensor transistor output circuits are either NPN (sinking) or ___ (sourcing).

T   F   **6.** For a 2-wire DC proximity sensor, the load is always in the circuit.

T   F   **7.** For a 2-wire AC proximity sensor, power is provided to the L2 (neutral) circuit and the load is in the line to the L1 (hot) terminal.

T   F   **8.** A cylindrical-style proximity sensor is designed with the sensing field at the end of the cylinder.

_____ **9.** ___ proximity sensors function by generating a pulse of sound waves.
A. Capacitance
B. Inductance
C. Ultraviolet
D. Ultrasonic

_____ **10.** Ultrasonic proximity sensors have a ___ close to the sensor where a target cannot be detected.
A. dead zone
B. hysteresis zone
C. frequency range
D. frequency band

_____ **11.** A diffused-mode photoelectric sensor detects light ___.
  A. reflected from the target
  B. reflected from a reflector
  C. with a separate receiver
  D. that becomes polarized

T   F   **12.** For a retro-reflective mode photoelectric sensor, the sensor is actuated when an object blocks the beam.

_____ **13.** A(n) ___ reflector has a triangular grooved surface that returns a light beam on a parallel axis.

SECTION
7 POSITION MEASUREMENT

chapter
28

Practical Position Measurement

REVIEW
QUESTIONS

Name _____   **Date** _____

_____ **1.** The ___ range of a proximity sensor is the range at which the target always actuates
the sensor.
    A. nominal
    B. actual
    C. working
    D. effective

T   F   **2.** The sensing range of a proximity sensor is increased if the target is constructed
of a material other than steel.

_____ **3.** Target objects spaced ___ each other may continuously activate a proximity
sensor.
    A. too close to
    B. too far apart from
    C. intermittently from
    D. at irregular intervals from

_____ **4.** ___ is the difference in range between the actuation point and the release point.
    A. Nominal range
    B. Frequency measurement
    C. Web guiding
    D. Hysteresis

T   F   **5.** A larger sensing range requires a larger coil.

_____ **6.** The ___ direction is the direction an object moves relative to a proximity
sensor.
    A. hysteresis
    B. actuation
    C. nominal
    D. actual

_____ **7.** ___ measure how often an object moves past the sensor.
    A. Mechanical switches
    B. Speed sensors
    C. Web guides
    D. Web loop controls

_____ **8.** Webs are handled with ___ that control the material through the process.
A. a series of rolls
B. mechanical switches
C. rotary sensors
D. light curtains

_____ **9.** A safety light curtain is used to ___ equipment if an operator reaches into a protected space.
A. detect damaged
B. start
C. shut down
D. all of the above

_____ **10.** With web loop control, it is common for ___ to be maintained between each driving roll.
A. take-up loops
B. thickness controllers
C. air nozzles
D. light curtains

Practical Position Measurement

Name _____   Date _____

Activity 28-1—Activating Frequency

Activating frequency is the limit to the number of pulses per second that can be detected by a photoelectric control in a given time period. All photoelectric controls have an activating frequency. To determine the required activating frequency of a photoelectric application, apply the following procedure:

1. Determine the maximum speed of the objects to be measured and convert that speed to the number of seconds it takes to travel 1″. For example, 21 ft/min = 0.238 sec/in.
2. Calculate the dark input signal duration. This is the time period when the photosensor is dark because the detected object is blocking the light beam. Dark input signal duration is found by multiplying the minimum dimension of the object to be detected (in in.) by the speed in sec/in. For example, if a 6″ × 3″ object with the 3″ dimension in the line of travel has a 0.263 sec/in. value, the dark input signal duration is 0.789 sec (3 × 0.263).
3. Calculate the light input signal duration. This is the time period when the photosensor is lit because no detectable object is in the light beam. Light input signal duration is found by multiplying the minimum distance between the objects to be detected by the speed in sec/in. For example, if the moving objects are spaced a minimum of 4.00″ apart on a conveyor traveling at 0.200 sec/in., the light input signal duration is 0.800 sec (4.00 × 0.200).
4. Calculate the activating frequency. This is the sum of the dark input and the light input signal durations. For example, if the dark input signal duration is 0.789 sec and the light input signal duration is 0.800 sec, the activating frequency is 1.589 sec.

A photoelectric sensor detects 2″ × 2″ objects that are 5″ apart and travel at 60 ft/min.

_____ **1.** What is the speed of the objects in sec/in.?

_____ **2.** What is the dark input signal duration?

_____ **3.** What is the light input signal duration?

_____ **4.** What is the activating frequency?

A photoelectric sensor detects 0.25″ square objects that are 0.05″ apart and travel at 150 ft/min.

_____ **5.** What is the speed of the objects in sec/in.?

_____ **6.** What is the dark input signal duration?

_____ **7.** What is the light input signal duration?

_____ **8.** What is the activating frequency?

Name _____   Date _____

Activity 28-2—Proximity Switch Mounting

Proximity sensors have a sensing head that produces a radiated sensing field. This sensing field detects the target. Interference with normal sensor operation can come from objects close to the sensor or from other sensors. General clearances are required for most proximity sensors. Spacing is measured from center to center of the sensors.

When flush mounting inductive and capacitive proximity sensors, a distance equal to or greater than twice the diameter of the sensors is required between sensors. If two sensors of different diameters are used, the diameter of the largest sensor is used for determining installation distances.

When non-flush mounting inductive and capacitive proximity sensors, a distance of three times the diameter of the sensor is required within or next to a material that can be detected. Three times the diameter of the largest sensor is required when inductive and capacitive proximity sensors are installed next to each other. When inductive and capacitive sensors are mounted opposite each other, six times the rated sensing distance is required for proper operation. When an inductive or capacitive sensor is mounted in a well or recess, the diameter of the recess must be at least three times the diameter of the sensor.

Use the illustration next to each question to answer the questions.

_____ **1.** What is the minimum distance required between the sensors for proper operation in mm?

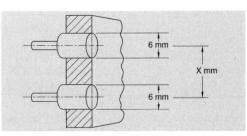

_____ **2.** What is the minimum distance required between the sensors and the surrounding material for proper operation in mm?

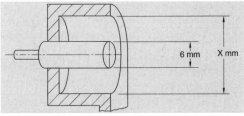

_____ **3.** What is the minimum distance required between the sensors for proper operation in mm?

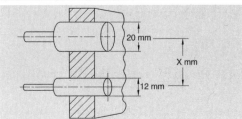

Name _____   **Date** _____

Troubleshooting Photoelectric Sensors

Troubleshooting circuits that include photoelectric switches requires an understanding of how the switch contacts operate in the circuit. If the photoelectric switch is set for dark operation, the switch's contacts change state (normally open close and normally closed open) any time there is an object in front of the eye (blocking light beam). If the photoelectric switch is set for light operation, the switch's contacts change state (normally open close and normally closed open) any time there is not an object in front of the eye (light beam not blocked).

In many applications, the photoelectric switch contacts are used to control a timer so that there is a time delay produced in the circuit. The timer contacts are then used to energize or de-energize a load. Thus, when troubleshooting an application that uses a timer with a photoelectric sensor, both the operation of the photo-electric sensor (light operated or dark operated) and the type of timer (on-delay, off-delay, one-shot, or recycle) must be understood. The photoelectric switch signals are used as inputs to timers, relays, counters, PLCs, and motor control circuits.

When troubleshooting conveyor systems containing photoelectric controls, approximate meter readings should be anticipated if the meter readings are going to be used to help determine circuit problems. Use the circuit on the back of this page to determine the expected DMM readings if the circuit is operating properly.

_____ **1.** The expected reading of DMM 1 with the conveyor ON without boxes of goods on the conveyor belt is ___ VAC.

_____ **2.** The expected reading of DMM 1 with the conveyor ON with boxes of goods on the conveyor belt backed up to the photo 2 position is ___ VAC.

_____ **3.** The expected reading of DMM 2 with the conveyor ON without boxes of goods on the conveyor belt is ___ VAC.

_____ **4.** The expected reading of DMM 2 with the conveyor ON with boxes of goods on the conveyor belt backed up to the photo 2 position is ___ VAC.

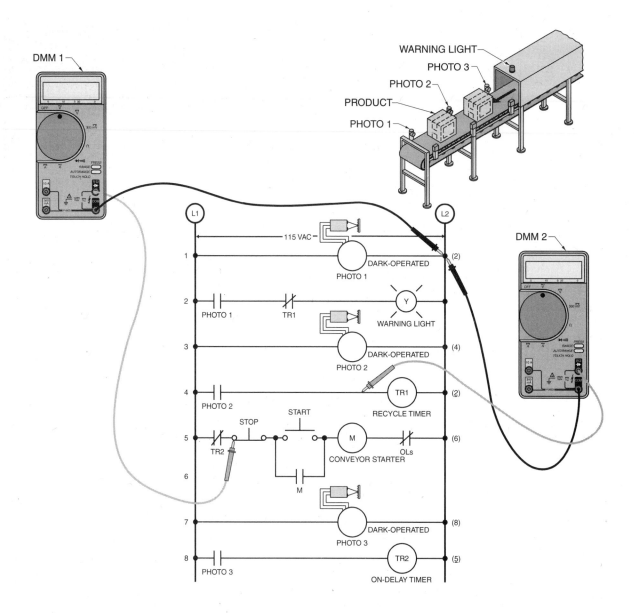

SECTION
8 TRANSMISSION AND COMMUNICATION

chapter
29
Transmission Signals

REVIEW
QUESTIONS

Name _____ **Date** _____

T   F   **1.** Voltage transmission is an electrical transmission method where the frequency of the signal is proportional to the measured value.

_____ **2.** ___ is a standardized method of conveying information from one instrument device to another.

_____ **3.** ___ is a series of discontinuous ON/OFF signals transmitted electrically.

_____ **4.** A(n) ___ is the data sent from one device to another by a specific method.
     A. digital number
     B. analog number
     C. transmission signal
     D. all of the above

_____ **5.** ___ data is a continuous range of values from a minimum to a maximum that can be related to a transmission signal range.
     A. Analog
     B. Digital
     C. Discrete
     D. Binary

_____ **6.** ___ transmission is an electrical transmission method consisting of a pure audible tone where the duration of the tone is proportional to the measurement value.
     A. Frequency
     B. Pulse
     C. Tone
     D. Voltage

T   F   **7.** Current transmission is an electric transmission system where a transmitting unit regulates resistance in a transmission loop.

T   F   **8.** Pulse transmission is an electric transmission method consisting of a rapid change in voltage from a low value to a high value and then back to the low value.

T   F   **9.** Frequency transmission is an electric transmission system where a transmitted voltage is the variable of interest.

_____ **10.** With a positive-displacement flowmeter, each pulse is calibrated to represent the ___.
     A. incremental flow quantity
     B. total flow over a set time
     C. continuous total flow
     D. all of the above

Name _____ Date _____

Activity 29-1—Analog Electric Signal Conversions

It is very useful to be able to quickly convert electric transmission signals to either a fractional value or percent and vice versa. Electric systems use a 4 mA to 20 mA signal range.

_____ 1. What is the percent signal corresponding to an 11.3 mA signal?

_____ 2. What is the current signal in mA corresponding to a signal that is 70% of full range?

_____ 3. What is the percent signal corresponding to a 5.6 mA signal?

_____ 4. What is the current signal in mA corresponding to a fraction of full scale of 0.333?

_____ 5. What is the current signal in mA corresponding to a fraction of full scale of 0.86?

_____ 6. What is the percent signal corresponding to an 18.5 mA signal?

_____ 7. What is the current signal in mA corresponding to a signal that is 56% of full range?

_____ 8. What is the current signal in mA corresponding to a signal that is 37% of full range?

_____ 9. What is the percent signal corresponding to a 4.8 mA signal?

_____ 10. What is the current signal in mA corresponding to a fraction of full scale of 0.95?

_____ 11. What is the percent signal corresponding to a 15.3 mA signal?

159

Name _____ **Date** _____

Activity 29-2—Transmission and Measurement Values

The transmission value, measurement value, transmission range, and measurement range are all related. When any three of the values are known, the fourth value can be calculated. These calculations are commonly done during the calibration of a transmitter or receiver.

Calculate the transmission value in mA. The transmission range is 4 mA to 20 mA.

		Measurement Range	Measurement Value
_____	**1.**	0°C to 100°C	41°C
_____	**2.**	0 scfh to 250 scfh	192 scfh
_____	**3.**	0% to 100% level	37%
_____	**4.**	0″ Hg vac to −30″ Hg vac	−20″ Hg vac
_____	**5.**	50°F to 250°F	143°F

Calculate the measurement value in the same units as the measurement range. The transmission range is 4 mA to 20 mA.

		Measurement Range	Transmission Value
_____	**6.**	0°C to 300°C	18.8 mA
_____	**7.**	50°F to 250°F	9.3 mA
_____	**8.**	0 gpm to 30 gpm	15.2 mA
_____	**9.**	0 psia to 30 psia	10.3 mA
_____	**10.**	0″ WC to 150″ WC	17.7 mA

Name _____ Date _____

Activity 29-3—Current to Voltage Conversions

Current transmission signals are converted to voltage signals by passing the 4 mA to 20 mA signal through a precision resistor (0.1% error). The resistor is usually 250 Ω but in special cases can be a different value.

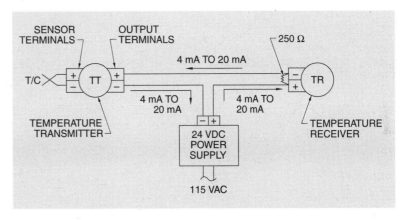

Given the following percent output signals, calculate the current signal in mA and the voltage drop in mV across a 250 Ω resistor.

_____ **1.** Given a percent output signal of 0%, the current output is ___ mA.

_____ **2.** Given a percent output signal of 0%, the voltage drop is ___ V.

_____ **3.** Given a percent output signal of 20%, the current output is ___ mA.

_____ **4.** Given a percent output signal of 20%, the voltage drop is ___ V.

_____ **5.** Given a percent output signal of 25%, the current output is ___ mA.

_____ **6.** Given a percent output signal of 25%, the voltage drop is ___ V.

_____ **7.** Given a percent output signal of 40%, the current output is ___ mA.

_____ **8.** Given a percent output signal of 40%, the voltage drop is ___ V.

_____ **9.** Given a percent output signal of 50%, the current output is ___ mA.

_____ **10.** Given a percent output signal of 50%, the voltage drop is ___ V.

_____ **11.** Given a percent output signal of 60%, the current output is ___ mA.

_____ **12.** Given a percent output signal of 60%, the voltage drop is ___ V.

_____ **13.** Given a percent output signal of 75%, the current output is ___ mA.

_____ **14.** Given a percent output signal of 75%, the voltage drop is ___ V.

SECTION
8 TRANSMISSION AND COMMUNICATION

chapter
30

Digital Numbering
Systems and Codes

REVIEW
QUESTIONS

Name _____   **Date** _____

_____ **1.** A(n) ___ is a group of 16 bits.

_____ **2.** A(n) ___ number is given in a base of 10.
      A. decimal
      B. octal
      C. hexadecimal
      D. binary

_____ **3.** The binary number 1010 is equal to the decimal number ___.

_____ **4.** A(n) ___ is a method of coding information in terms of two-position ON/OFF voltage signals.

   T   F **5.** A byte is a grouping of four bits and is the minimum number of bits needed to represent a single decimal number.

_____ **6.** A(n) ___ system is a numbering system that uses groups of four bits to represent a group of decimal numbers.
      A. binary coded decimal (BCD)
      B. octal
      C. hexadecimal
      D. nibble

_____ **7.** A(n) ___ number is given in a base of 2.
      A. octal
      B. decimal
      C. binary
      D. hexadecimal

   T   F **8.** A nibble is a group of eight bits.

   T   F **9.** The largest-valued symbol always has a value of one less than the base.

_____ **10.** In binary, usually 0 is called a ___ state and 1 is called a ___ state.
      A. positive; null
      B. null; positive
      C. true; false
      D. false; true

_____ **11.** A(n) ___ signal is a group of low-level DC voltage pulses that can be used to convey information.
A. analog
B. digital
C. pneumatic
D. tone

_____ **12.** A(n) ___ number is given in a base of 16.

_____ **13.** A(n) ___ is used to distinguish the base used when there is the possibility of confusion between different numbering systems.

_____ **14.** 101B is a(n) ___ number.
A. binary
B. decimal
C. hexadecimal
D. octal

_____ **15.** A(n) ___ code is a number code that uses a combination of letters, symbols, and decimal numbers to process information for computers and printers.

_____ **16.** An octal number is given in a base of ___.
A. 2
B. 4
C. 8
D. 10

_____ **17.** ___ is a seven-bit or eight-bit alphanumeric code used to represent the basic alphabet plus numbers and special symbols.

T   F   **18.** A binary number can be either signed or unsigned.

_____ **19.** The decimal number 14 is equal to the binary number ___.

_____ **20.** A(n) ___ is a binary digit consisting of a 0 or a 1.

SECTION
8 TRANSMISSION AND COMMUNICATION

Digital Numbering
Systems and Codes

chapter
30
ACTIVITIES

Name _____ Date _____

Activity 30-1—Number Conversions

Convert the decimal numbers into the equivalent format requested.

		Decimal Number	Convert to
_____	1.	4666	16-bit binary
_____	2.	4666	hexadecimal
_____	3.	4666	BCD
_____	4.	4666	octal
_____	5.	41,321	16-bit binary

Convert the 16-bit string of 1s and 0s into the numerical format requested.

		16-Bit String	Convert to
_____	6.	0010 0011 1011 0001	decimal
_____	7.	1001 0111 0101 0110 BCD	decimal
_____	8.	0011 1110 1011 0011	hexadecimal
_____	9.	0110 0011 1100 1001	decimal
_____	10.	1001 0011 1001 0111 BCD	decimal

SECTION
8 TRANSMISSION AND COMMUNICATION

chapter
31

Digital Communications

REVIEW
QUESTIONS

Name _____   **Date** _____

_____ 1. A(n) ___ wiring format is a method of communications wiring where bits of information are transferred one after another over a pair or pairs of wires.
   A. series
   B. parallel
   C. series/parallel
   D. analog

_____ 2. ___ is the method of sending and receiving binary information, in the form of low-level DC voltage pulses, between multiple devices using a common wiring format and procedure.

_____ 3. A(n) ___ network is a communications system that allows distributed control systems (DCSs), PLCs, or other controllers to communicate with I/O devices and each other.

_____ 4. ___ communication is a serial communications system developed by the Electronics Industry Association and the Telecommunications Industry Association.
   A. RJ-444
   B. RJ-422
   C. RS-323
   D. RS-232

_____ 5. A ___ is a data transmission path where digital signals are dropped off or picked up at each point where a device is attached to wires.
   A. bus
   B. DCS
   C. PLC
   D. serial address

_____ 6. A(n) ___ is a unique number assigned to each device on a network.
   A. ID position
   B. network address
   C. service address
   D. parallel address

_____ 7. Transistor-___ logic is a connection method that allows separate solid-state devices to be wired together to pass information.
   A. transistor
   B. resistor
   C. thermistor
   D. none of the above

T   F   **8.** In optical fiber communications, digital electronic communications are converted to a solid-state laser light-transmitted signal.

_____ **9.** A ___ circuit is a circuit in a transmitting device that takes the positive voltage generated by the receiving device and shunts it to ground, lowering the voltage at the receiver circuit.
     A. transmitter
     B. receiver
     C. sink
     D. source

_____ **10.** A(n) ___ wiring format is a method of communications wiring where each bit of information is transferred on its own dedicated wire.
     A. RS-232
     B. consecutive
     C. series
     D. parallel

_____ **11.** A(n) ___ is a pair of twisted wires with no electromagnetic shielding.
     A. unshielded twisted pair
     B. EMI pair
     C. series wiring connection
     D. parallel wiring connection

Name _____ Date _____

Activity 31-1—Network Configurations

Draw a diagram showing how controllers are connected in the following network configurations.

1. Star

2. Ring

3. Bus

Digital Communications

Name ______________________________ Date ______________

Activity 31-2—Digital Circuits

1. What special electrical device is used in TTL circuits to electrically isolate the transmitter from the receiver?

___

___

___

2. Explain what a sinking circuit does to pass digital information to the next device.

___

___

___

3. Explain what a sourcing circuit does to pass digital information to the next device.

___

___

___

Name _____ **Date** _____

Activity 31-3—Series Circuits

The three types of digital series communication are simplex, half-duplex, and full-duplex.

1. Draw a simplex digital circuit using a single line to indicate the direction of communication.

2. Draw a half-duplex digital circuit using a single line to indicate the direction of communication.

3. Draw a full-duplex digital circuit using a single line to indicate the direction of communication.

Name _____   **Date** _____

Activity 31-4—Digital Wiring

1. Draw an RS-232 wiring circuit between two devices.

2. Draw an RS-485 wiring circuit between two devices.

172

SECTION
8 *TRANSMISSION AND COMMUNICATION*

Industrial Networks

chapter
32

REVIEW
QUESTIONS

Name _____   **Date** _____

T   F   **1.** Profibus is a process bus network capable of communicating information between a master controller and an intelligent slave process field device.

_____ **2.** A ___ is a unit of data sent across a network.
　　　　　A. bit
　　　　　B. nibble
　　　　　C. byte
　　　　　D. packet

_____ **3.** ___ is a high-speed, interference-immune communications method using the Transmission Control Protocol/Internet Protocol (TCP/IP) format.
　　　　　A. Centronics
　　　　　B. Ethernet
　　　　　C. ASCII
　　　　　D. Profibus

_____ **4.** The ___ protocol is a hybrid communications system combining digital and analog communication on the same wire.
　　　　　A. Profibus-FMS
　　　　　B. Profibus-DP
　　　　　C. HART®
　　　　　D. MODBUS

_____ **5.** An ___ connector is a snap-in connector that looks like a large phone plug and contains eight pins instead of four as in standard phone connectors.
　　　　　A. M-12
　　　　　B. RJ-45
　　　　　C. M-45
　　　　　D. RS-232

_____ **6.** A fieldbus is an open, digital, ___, two-way communication network that connects with high-level information devices.
　　　　　A. serial
　　　　　B. parallel
　　　　　C. analog
　　　　　D. physical

_____ **7.** A 2-wire fieldbus network uses ___ to furnish power to field instruments.
      A. two separate power wires
      B. field power lines
      C. communication wires
      D. baseband communication

_____ **8.** Networks can be broadly classified as sensor bus (bit), device bus (byte), and ___ networks.
      A. fieldbus (message)
      B. series (sequential)
      C. parallel (duplicate)
      D. series/parallel

_____ **9.** Category 5 unshielded twisted pair consists of ___ twisted pairs of copper wire.
      A. two
      B. three
      C. four
      D. eight

T   F   **10.** The primary difference between FOUNDATION and other fieldbus systems is that the software contains function blocks.

Name _____ Date _____

Activity 32-1—Fieldbus

A conventional DCS or PLC control system typically has separate wire pairs running from the controller to the individual field devices. There are many different fieldbus systems with a wide range in capabilities, but they all have one thing in common. Fieldbus systems have a single communication cable between the controller and the field instruments.

A completed drawing of the conventional DCS system is shown on the left. The drawing of the fieldbus system on the right is incomplete.

1. Complete the drawing of the fieldbus system showing how the field devices are wired to the controller.

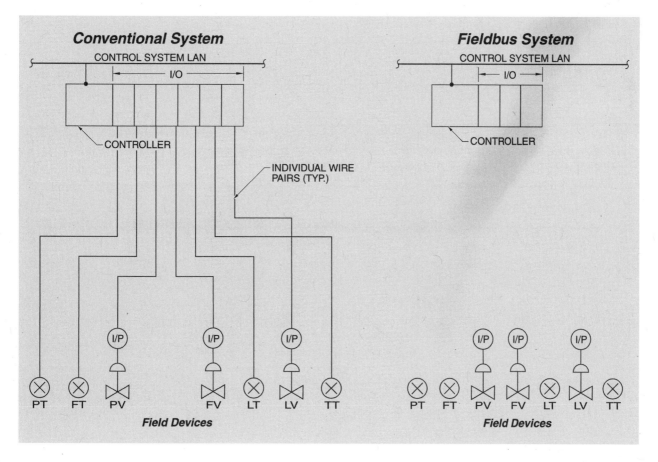

Name _____   **Date** _____

Activity 32-2—Digital Wiring

1. There are two different wiring formats for Ethernet circuits: "AT&T" or "T268B"; and "EIA" or "T568A." Explain the differences.

   _____

   _____

2. When are Ethernet straight-through wiring cables used and when are crossover cables required?

   _____

   _____

3. What type of connectors are used for Ethernet cables?

   _____

   _____

SECTION
8 TRANSMISSION AND COMMUNICATION

chapter
33

Wireless Systems

REVIEW
QUESTIONS

Name _____ **Date** _____

_____ **1.** Wireless transmission is the method of using ___ frequencies to transfer information from one device to another.
    A. medium
    B. wavelength
    C. radio
    D. Ethernet

_____ **2.** ___ is a transmission method that uses multiple frequencies to transmit data.
    A. Single channel radio
    B. Spread spectrum
    C. Unidirectional antenna gain
    D. Yagi gain

_____ **3.** WPA is a security standard with advancements such as authentication and ___.
    A. ISM
    B. dynamic keys
    C. mesh networking
    D. PSK

_____ **4.** At a lower bit rate, there is ___ transmission through barriers than at a higher bit rate.
    A. better
    B. worse
    C. negligible
    D. decreased

_____ **5.** ___ is a strategy a device uses for operation whereby it obtains power from its surrounding environment.
    A. Battery charging
    B. Battery storage
    C. Energy harvesting
    D. Sleeping

_____ **6.** The ability of devices to relay signals for other devices to reduce the radio power output needed is called ___.
    A. power transfer
    B. interference modeling
    C. bit rate control
    D. mesh networking

_____ **7.** A(n) ___ antenna is used to focus transmission energy in one direction.
       A. unidirectional
       B. omnidirectional
       C. mesh
       D. spread spectrum

_____ **8.** ___ is a wireless communications protocol designed for monitoring and control for distances up to 75 m with up to 250 nodes per network.
       A. Zigbee®
       B. Bluetooth®
       C. Mesh networking
       D. Spread spectrum

_____ **9.** A(n) ___ is the component of a wireless system that encrypts the information being sent using a continuously changing key.
       A. WEP control
       B. dynamic key
       C. ISM link
       D. Yagi modulator

T   F   **10.** General packet radio service (GPRS) communications is a voice radio service for wireless networks.

SECTION
8 TRANSMISSION AND COMMUNICATION

chapter
34

Practical Transmission
and Communication

REVIEW
QUESTIONS

Name _____   **Date** _____

_____ 1. A ___ is a signal transmission cable in which two conductors carry the signal and create a balanced circuit alternating from positive to negative.
A. Cat 5 cable
B. Cat 6 cable
C. coaxial line
D. balanced line

_____ 2. A(n) ___ is the current flow from one grounded point to a second grounded point in the same powered loop.
A. ground loop
B. balanced line
C. isolated device
D. current repeater

_____ 3. A properly balanced unshielded twisted pair provides significant immunity to ___.
A. ground loops
B. current impedance
C. electromagnetic interference (EMI)
D. dynamic keys

_____ 4. ___ is important because a loop with too much impedance is not able to generate enough current to deliver the 20 mA that is necessary for signaling.
A. Allowable loop impedance
B. Cable shielding
C. Electromagnetic interference (EMI)
D. Ground looping

_____ 5. A ___ can be used when a loop cannot generate enough loop current.
A. voltage source
B. current repeater
C. balanced line
D. shielded cable

_____ 6. A(n) ___ can be caused when there is more than one ground point for multiple loops that are connected together.
A. false instrument reading
B. ground loop
C. leakage current
D. all of the above

_____ **7.** A(n) ___ connector is a 4-wire or 8-wire connector with threaded metal fittings designed for an industrial environment.
    A. M-12
    B. RJ-45
    C. UTP
    D. EMI

_____ **8.** A(n) ___ can be used as a possible solution to EMI problems.
    A. shielded cable
    B. optical fiber cable
    C. balanced line
    D. all of the above

_____ **9.** A ___-wire transmitter uses the 24 VDC power from a current transmission loop to power the transmitter.
    A. single
    B. 2
    C. 3
    D. 4

_____ **10.** A smart transmitter ___ so that it can handle multiple inputs and outputs, communicate and change configuration details, and signal alarms and error conditions.
    A. combines digital and analog signals
    B. uses an analog signal
    C. uses a balanced line
    D. uses shielded cable

SECTION
8 TRANSMISSION AND COMMUNICATION

chapter
34

Practical Transmission
and Communication

ACTIVITIES

Name _____ **Date** _____

Activity 34-1—Allowable Loop Impedance

A temperature transmitter can handle a total impedance of 650 Ω in its external circuits when powered by a 24 VDC power supply. The transmitter is in a control loop circuit and sends a 4 mA to 20 mA signal to a stand-alone digital controller. A current trip switch uses a 250 Ω resistor on the input and has relay contacts that can be set to trip at a predetermined signal value. The measurement circuit is powered by a separate isolated 24 VDC power supply. The controller output goes to an I/P converter mounted on a fail closed control valve. The output is powered from the controller's internal power supply.

1. Draw the circuit as described above.

2. It is decided to record the above temperature transmitter output. Draw the additional circuits necessary to do this without using a current rectifier.

Name _____ **Date** _____

Loop Power Supplies

There are three identical circuits. In each circuit, a temperature transmitter has an allowable total impedance of 650 Ω in its external circuits when powered by a 24 VDC power supply. The transmitter is in a control loop circuit and sends a 4 mA to 20 mA signal to a stand-alone digital controller. A current trip switch uses a 250 Ω resistor on the input and has relay contacts that can be set to trip at a predetermined signal value.

1. Draw the three control system measurement circuits as they would need to be configured to be powered by a single isolated 24 VDC power supply. Show the location of the 24 VDC power fused disconnects.

SECTION
9 AUTOMATIC CONTROL

chapter
35

Automatic Control and
Process Dynamics

REVIEW
QUESTIONS

Name _____   **Date** _____

T   F   **1.** Impedance is the opposition to the potential that moves material or energy in or out of a process.

_____   **2.** An ___ reaction is a chemical reaction that generates heat during the reaction and increases temperature.
       A. endothermic
       B. exothermic
       C. enthalpic
       D. entropic

T   F   **3.** Centrifugal acceleration is the driving force that causes material or energy to move through a process.

_____   **4.** An ___ reaction is a chemical reaction that consumes heat, and more heat energy must be added to sustain the reaction.
       A. enthalpic
       B. entropic
       C. endothermic
       D. exothermic

T   F   **5.** A step change is a sudden change in an input variable in a process that is managed by a controller.

_____   **6.** Process dynamics are the attributes of a process that describe how a process responds to ___ changes imposed upon it.

T   F   **7.** A time constant (τ) is the measured time when an output response is 1.00 or 100% of an input step change.

_____   **8.** A ___ change is an uncontrolled change in the process operating conditions that changes the PV and must be compensated for by a change in the CV.
       A. setpoint
       B. static
       C. temperature
       D. load

_____   **9.** ___ is the equipment and techniques used to automatically regulate a process to maintain a desired outcome.

_____   **10.** ___ is the ratio of the change in output to the change in input of a process.

_____11. A(n) ___ curve is a plot of the PV against the CV.
        A. open loop
        B. process
        C. closed loop
        D. gain

_____ 12. A ___ variable is an independent variable in a process control system that is used to adjust a dependent variable.
        A. control
        B. setpoint
        C. dynamic
        D. controller

_____ 13. A(n) ___ process is a process where the gain at any point of the input range is the same as the gain at any other point.

T   F   14. Transmitter dynamic gain is the amount of output change from a transmitter for a specific input change.

_____ 15. A(n) ___ process is a process where the gain changes at different points on a process curve.

_____ 16. A(n) ___ is a device that compares a process measurement to a setpoint and changes the CV to bring the PV back to the setpoint.

T   F   17. Dead time is a delay in the response of a process that represents the time it takes for a process to respond completely when there is a change in the inputs to the process.

_____ 18. A primary element is the sensing device that detects the condition of a ___ variable.
        A. control
        B. process
        C. setpoint
        D. secondary

T   F   19. Capacitance is the ability of a process to store material or energy.

_____ 20. A(n) ___ is a device that receives a control signal and regulates the amount of material or energy in a process.

T   F   21. Lag time is the period of time that occurs between the time a change is made to a process and the time the first response to that change is detected.

_____ 22. A ___ variable is a dependent variable that is controlled in a control system.
        A. manipulated
        B. control
        C. process
        D. static

_____ 23. ___ action is the situation where the controller output increases with an increase in the measurement of the PV.

T   F   24. Offset is a steady-state error that is a permanent part of a system.

_____ 25. ___ transfer is a controller function included so that there is no sudden change in output value when a controller is switched from automatic to manual mode or back again.
      A. Bumpless
      B. Static
      C. Output
      D. Automatic

T   F   **26.** Feedback is a rarely used control function.

_____ 27. A(n) ___ is the logic and wiring for a transfer of information from a primary element to a controller, from the controller to a final element, from the final element to a process, and from the process back to the primary element.

T   F   **28.** Valves are frequently configured to be fail closed for safety reasons.

_____ 29. A(n) ___ loop is a control system that sends a control signal to a final element but does not verify the results of that control.

_____ 30. A(n) ___ is the desired value at which a process should be controlled and is used by a controller as a reference for comparison with the PV.

_____ 31. An open loop process response curve is a graph of the results of a(n) ___ in a manually adjusted output signal of an open loop controller that results in a change in a process measurement.
      A. break
      B. offset
      C. step change
      D. error

T   F   **32.** All controllers must have a setpoint to be able to provide a control function.

T   F   **33.** Error is the difference between a process variable and a setpoint.

_____ 34. ___ action is the situation where the controller output decreases with an increase in the measurement of the PV.

T   F   **35.** A control loop is a control design where a controller is connected to a process in an arrangement such that any change in the process is measured and used to adjust action by the controller.

T   F   **36.** A typical thermostat in an HVAC system uses feedforward control.

_____ 37. A(n) ___ is a control system that provides feedback to a controller on the state of a process variable.
      A. control loop
      B. closed loop
      C. setpoint
      D. offset

_____ 38. Setpoint tracking is the technique of storing the ___ in a setpoint memory module while the controller is in manual.
      A. control variable
      B. setpoint
      C. process variable
      D. manipulated variable

Name _____   Date _____

Activity 35-1—Gravity Tank Outflow

A tank has a gravity flow outlet controlled by a manual valve. The level control system measures the level and controls the feed to the tank with a throttling control valve. When the height in the tank is 10 ft, the outlet flow is 30 gpm. The basic equation that relates the flow to the height of the liquid in the tank is a follows:

$$F = C \times \sqrt{h}$$
where
F = flow
C = characteristic constant
h = height

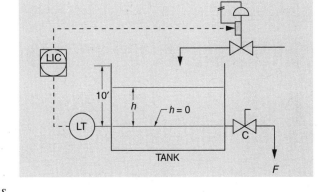

Use the illustration to complete the following activities.

1. Calculate the characteristic constant, C, for this situation.

   _____

2. Plot the flow on the horizontal axis and the height on the vertical axis for flows up to 30 gpm.

3. Calculate the process gain at 2.5′ as a change in flow per change in height and as a dimensionless gain in percent.

   _____

4. Calculate the process gain at 5′ as a change in flow per change in height and as a dimensionless gain in percent.

   _____

5. Calculate the process gain at 7.5′ as a change in flow per change in height and as a dimensionless gain in percent.

   _____

6. Is the process linear or nonlinear. Why?

   _____

Name _____ Date _____

Activity 35-2—Pumped Tank Outflow

A tank has a pumped outlet flow controlled by a manual valve. The level control system measures the level and controls the feed to the tank through a throttling control valve. The flow out of the tank and change of height of the liquid are as follows:

$$F_{out} = C \text{ and } \Delta h = (F_{in} - F_{out}) \times K$$

where
F_{out} = flow out of the tank, in gpm
F_{in} = flow into the tank, in gpm
C = outflow coefficient, in gpm
Δh = change in height, in ft
K = tank coefficient, in gal./ft

Use the illustration to answer the following questions.

1. Is there an equation that relates the outflow F_{out} to the height in the tank?

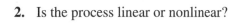

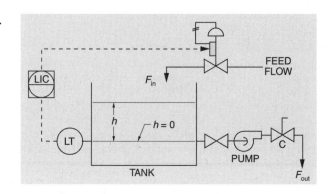

2. Is the process linear or nonlinear?

_____

3. How does a change in the value of the outflow coefficient affect the linearity of the process?

_____

4. What is another term for the outflow coefficient when considering process dynamics?

_____

SECTION
9 AUTOMATIC CONTROL

Automatic Control and
Process Dynamics

chapter
35
ACTIVITIES

Name _____   **Date** _____

Activity 35-3—First-Order Response

A tank level control system has a steady level measurement value of 47% on a scale of 0% to 100%. The controller is switched to manual, the controller output is changed from 52% to 60%, and the measured level starts changing. The level change over time is recorded and shown here. The first-order response equation is as follows:

output = initial value + step change $\times (1 - e^{-t/\tau})$

t (min)	LEVEL (%)	t (min)	LEVEL (%)
0.0	47.0	5.0	57.2
0.5	48.8	5.5	57.6
1.0	50.4	6.0	57.9
1.5	51.8	7.0	58.5
2.0	53.0	8.0	58.9
2.5	54.0	9.0	59.2
3.0	54.8	10.0	59.4
3.5	55.6	11.0	59.6
4.0	56.2	12.0	59.7
4.5	56.7	15.0	59.9

Use the graph paper on the back of this page and the level table to answer the following questions.

1. Plot the process response level data on the graph paper on the back of this page for time from 0 min to 15 min and level from 47% to 60%.

2. From the plot of the process response data, determine the first-order time constant.

   _____

3. Use the first-order response equation to calculate the output at time equal to 1 time constant.

   _____

4. Does this process follow a first-order response?

   _____

5. What is the gain of this process response?

   _____

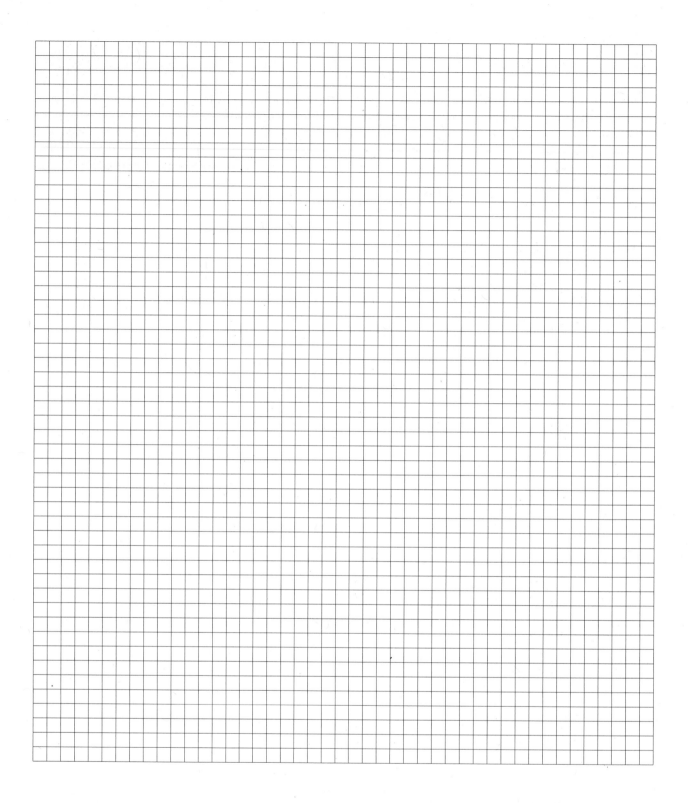

Automatic Control and
Process Dynamics
ACTIVITIES

Name _____ Date _____

Activity 35-4—Open Loop Transient Response

A steam-heated jacketed reactor has a temperature controller. An open loop transient response test is conducted. The results of the test are shown.

TIME (min)	TEMP. (°F)	TIME (min)	TEMP. (°F)
0	141.0	11	150.5
1	141.0	12	150.8
2	141.0	13	151.1
3	141.2	14	151.2
4	141.3	15	151.3
5	142.8	16	151.35
6	145.2	17	151.4
7	147.0	18	151.45
8	148.5	19	151.45
9	149.3	20	151.45
10	150.0		

Use the graph paper on the back of this page and the table shown above to answer the following questions.

1. Plot the process response data provided in the table for time from 0 min to 20 min.

2. Is this response a first-order curve? Explain.

   _____

3. What is the dead time?

   _____

4. What is the largest lag time constant?

   _____

5. What is the percentage of dead time compared to the largest lag time constant? What does this mean?

   _____

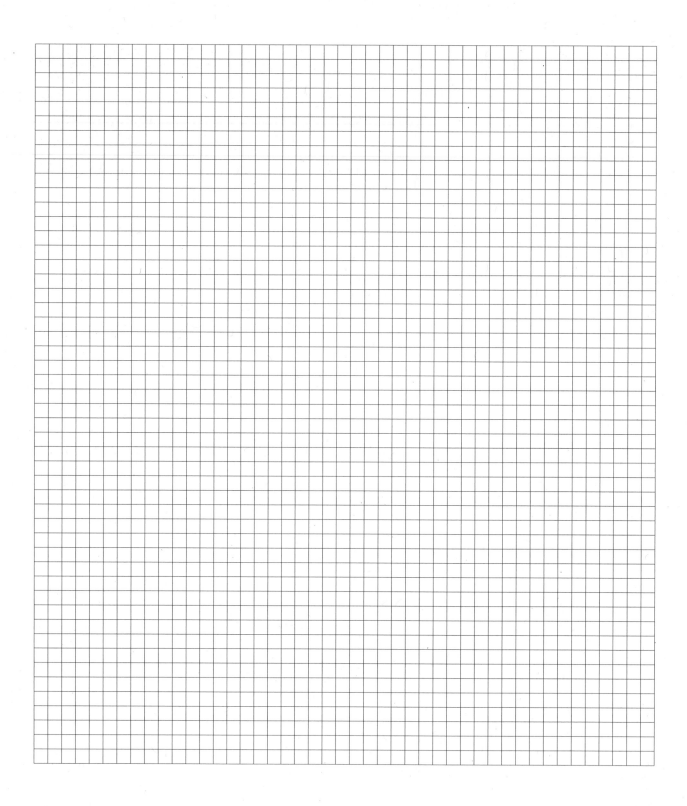

SECTION
9 AUTOMATIC CONTROL

Automatic Control and
Process Dynamics

chapter
35

ACTIVITIES

Name _____   **Date** _____

Activity 35-5—Controller Actions

Use the description and the illustration to determine whether the controller should use direct or reverse action.

_____ **1.** A flow control with a fail closed (air to open) control valve.

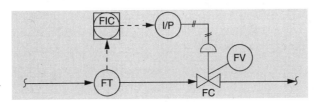

_____ **2.** A level-controlled tank with a fail open (air to close) control valve in the feed line to the tank.

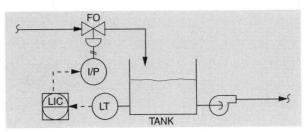

_____ **3.** A pressure control system with a downstream pressure controlled by a fail closed (air to open) control valve.

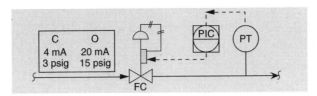

195

_____ 4. A temperature control system controlling the process fluid temperature from a steam-heated exchanger using a fail closed (air to open) control valve.

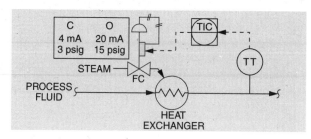

_____ 5. A backpressure control system with the downstream gas going to two possible areas. One area is an atmospheric vent through a fail open (air to close) control valve and the other is a fail closed (air to open) control valve to a compressor inlet.

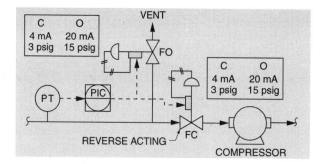

Name _____ **Date** _____

_____ **1.** A(n) ___ control system is a control system with multiple loops where a primary variable is controlled by changing the setpoint of a related secondary controller.

_____ **2.** A heat ___ is a small electric heater that is part of a thermostat.
   A. anticipator
   B. adjuster
   C. controller
   D. integrator

T   F **3.** Adaptive gain control is a control strategy where a nonlinear gain can be applied to a nonlinear process.

T   F **4.** Integral time (T_I) is the time it takes for a controller to change a CV by 50% for a 10% change in the difference between CV and SP.

_____ **5.** A(n) ___ output is an output that has a continuous range of possible output values between minimum and maximum limits.
   A. digital
   B. discrete
   C. analog
   D. binary

T   F **6.** Proportional gain is the gain, or sensitivity, of a control variable.

_____ **7.** ___ is the integral time multiplied by controller gain.

T   F **8.** A proportional (P) control strategy is a method of changing the output of a controller by an amount proportional to the error.

_____ **9.** ___ control is proportional control combined with integral control.

_____ **10.** ___ is the range of values of an input that corresponds to a full range of output from a controller, stated as a percentage.
   A. Proportional band
   B. Throttling range
   C. Proportional gain
   D. Integral gain

T   F **11.** ON/OFF control is a method that produces an output that provides only an ON or OFF signal to the final element of the process.

_____ **12.** ___ is a controller function that positions a final element in a central position when the process variable is at setpoint.
   A. Overshoot
   B. Integral gain
   C. Proportional gain
   D. Output bias

198 INSTRUMENTATION WORKBOOK

_____ **13.** ___ is the change of the PV that exceeds the upper deadband value when there is a disturbance to the system.
   A. Undershoot
   B. Gain
   C. Overshoot
   D. Tuning

_____ **14.** An integral (I) control strategy is a method of changing the ___ of a controller by an amount proportional to the error and the duration of that error.
   A. integral gain
   B. output
   C. proportional gain
   D. derivative gain

_____ **15.** A(n) ___ controller is an ON/OFF controller that has a predetermined output period during which the output contact is held closed for a variable portion of the output period.

_____ **16.** ___ is the integral controller mode, so called because in the older proportional-only controllers the operator had to manually reset the setpoint.
   A. Integral gain
   B. Automatic reset
   C. Throttling range
   D. Overshoot

_____ **17.** ___ is the change of the PV that goes below the lower deadband value when there is a disturbance to the system.

T   F   **18.** The reset rate is the reciprocal of integral time.

_____ **19.** A(n) ___ control strategy is a method of changing the output of a controller in proportion to the rate of change of the process variable.

_____ **20.** ___ is the number of units of a process variable that causes an actuator to move through its entire range.

T   F   **21.** ON/OFF proportional control is proportional control combined with both integral control and derivative control.

_____ **22.** A(n) ___ is the range of values where a change in measurement value results in no change in controller output.
   A. proportional band
   B. lag
   C. open loop
   D. deadband

_____ **23.** ___ control is a control strategy that only controls the inputs to a process without feedback from the output of the process.
   A. Feedback
   B. Feedforward
   C. Proportional
   D. ON/OFF

_____ **24.** A(n) ___ response diagram is a curve that shows controller response to a given measurement without the controller actually being connected to the process.

Name _____   Date _____

Activity 36-1—ON/OFF Control

The initial slope of a first-order response curve intersects the final value line at a time equal to the time constant of the first-order response. This means that, for a process with a first-order response that is forced to change at a defined rate from an initial measurement to a final measurement, the process time constant can be determined by knowing the final possible high and low values.

An ON/OFF heating system provides hot air at a temperature of 100°F when the heating system is ON. With the heating system OFF, the minimum temperature is 40°F. The setpoint is 70°F. The heating time constant is the amount of time required to heat the process from the setpoint to the maximum possible temperature when heated at the initial rate of change. The cooling time constant is the time required for the process to cool down from the setpoint to the minimum temperature when the heating system is OFF and cooled at the initial rate of change.

The process has a 1°F deadband and stays ON for 2 min and stays OFF for 5 min.

1. What is the heating time constant?

   _____

2. What is the cooling time constant?

   _____

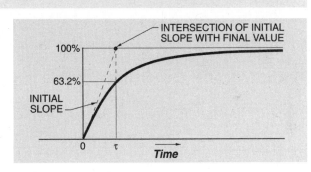

The process has a 2°F deadband, a dead time of 2 min, a heating time of 4 min to move through the dead-band, and a cooling time of 8 min to move through the deadband.

3. Draw the heating and cooling response curves on the graph paper on the back of this page.

4. What is the heating time constant?

   _____

5. What is the cooling time constant?

   _____

6. What is the amount of overshoot during the heating cycle?

   _____

7. What is the amount of undershoot during the cooling cycle?

   _____

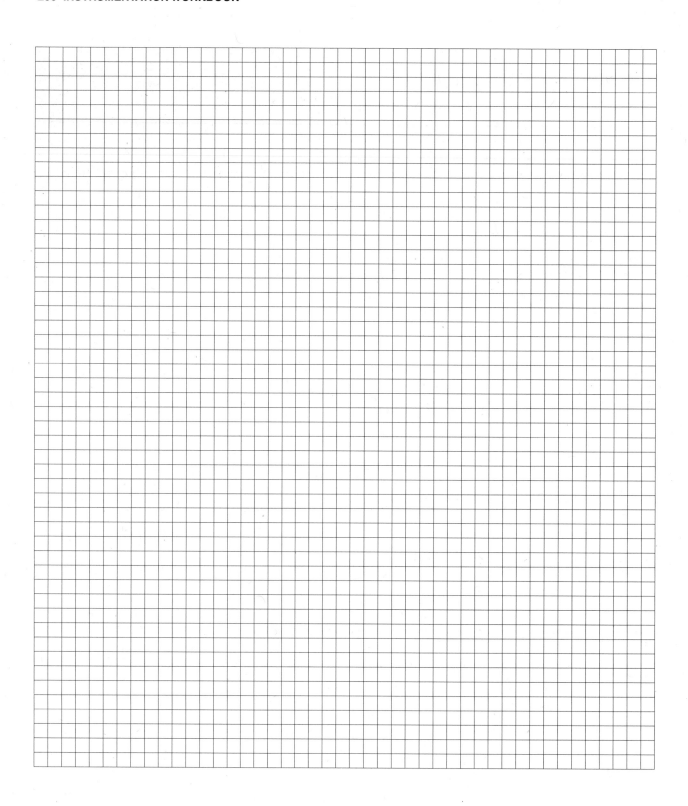

Name _____ **Date** _____

Activity 36-2—Proportional Control

A proportional only, direct-acting controller with a 50% output bias is set up with a simulated input for the measurement and an output meter. Both the simulated input and the output meter have a scale of 0% to 100%. The calculation of the output uses the following equation:

$$output = \frac{CA \times (PV - SP)}{MR} \times \frac{100}{PB} \times 100 + bias$$

where
CA = controller action
PV = process variable
SP = setpoint
MR = measurement range
PB = proportional band
bias = predetermined value for bias

Given the following conditions, calculate the output.

_____ 1. PB = 100%; setpoint = 50%; measurement = 50%

_____ 2. PB = 200%; setpoint = 75%; measurement = 25%

_____ 3. PB = 50%; setpoint = 25%; measurement = 50%

_____ 4. PB = 25%; setpoint = 50%; measurement = 60%

_____ 5. PB = 125%; setpoint = 40%; measurement = 50%

_____ 6. PB = 20%; setpoint = 45%; measurement = 50%

Name _____   Date _____

Activity 36-3—PI Open Loop Response

A proportional-integral, direct-acting controller is set up with a simulated measurement and a meter on the output. The measurement range is 0 to 200 and the output range is 0% to 100%. The PB is 75% and the integral time is 10 min/R. The initial output is 45%. The measurement is at the setpoint of 110. At time zero, the measurement is ramped up at 1 unit/min. After 20 min, the measurement stops increasing and holds steady for 10 min. At this point, the measurement returns to the setpoint with a step change.

1. Graph the measurement value from 0 min to 35 min.

2. Calculate the proportional output at 5 min intervals from 0 min to 35 min.

_____

3. Calculate the integral output at 5 min intervals from 0 min to 35 min.

_____

4. Calculate the total change in output and the total output at 5 min intervals from 0 min to 35 min.

_____

5. Graph the total output value from 0 min to 35 min.

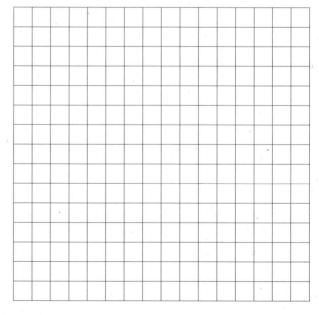

Name _____   Date _____

Activity 36-4—PD Open Loop Response

The original design of PD and PID controllers had the derivative function acting on the rate of change of the deviation (the difference between the measurement and the setpoint). Modern controllers have the derivative function acting only on the rate of change of the measurement. For the derivative to respond properly to measurement changes, there has to be a gain value associated with the derivative. In most controllers, the user does not have the ability to change this gain. If the gain is set to 1.0, the output response from the controller is the same as would be generated with the older style controller.

The derivative action on the measurement can be calculated for a ramped measurement change as follows:

$$\begin{pmatrix} \text{derivative action on} \\ \text{measurement change} \end{pmatrix} = \begin{pmatrix} \text{rate of change} \\ \text{of measurement} \end{pmatrix} \times \begin{pmatrix} \text{derivative} \\ \text{time} \end{pmatrix} \times \begin{pmatrix} \text{derivative} \\ \text{gain} \end{pmatrix}$$

A direct-acting PD (proportional-derivative) controller is set up with a simulated measurement input and a measured output. The initial measurement value is 80°F, measurement range is 0°F to 150°F, initial output is 40%, proportional band is 50% (gain = 2), derivative time is 20 min, and the derivative gain is 1.0. At time = 0, the measurement ramps up at a rate of 1°F/min. The setpoint is 80°F.

Use the above information to perform the following tasks.

1. Calculate the derivative action on the measurement value at 0 min and at 20 min for a modern controller.

   _____

2. Calculate the proportional response at 0 min and at 20 min for a modern controller.

   _____

3. Calculate the total output at 0 min and 20 min for a modern controller.

   _____

4. Graph the measurement and the total output values from 0 min to 20 min for a modern controller.

5. Calculate the proportional response at 0 min and 20 min for an older style controller.

_____

6. Calculate the derivative action at 0 min and 20 min for an older style controller.

_____

7. Calculate the total output at 0 min and 20 min for an older style controller.

_____

8. Graph the measurement and the total output values from 0 min to 20 min for an older style controller.

Name _____ **Date** _____

Activity 36-5—PID Open Loop Response

A proportional-derivative control output response is the same for controllers with the derivative acting on the measurement and with the derivative acting on the deviation. The output for a proportional-integral-derivative controller is not the same for the two different styles of derivative. This is because the derivative acting on the measurement affects the output response from both the proportional and integral actions.

A direct-acting PID controller has a setpoint of 50% and an output of 40%. The proportional band is 75% with an integral time of 10 min/R, a derivative time of 7.5 min, and a derivative gain of 1.0. At time = 0, the measurement ramps up at 1%/min for 10 min and then ramps down and takes 15 min to return to the setpoint. At the point, the measurement remains at the setpoint. The derivative acts on the measurement.

Although the output from the derivative action of both the modern and older style controllers is the same, the change in measurement also affects the integral response so that the total output is not the same.

The derivative action on the measurement can be calculated for a ramped measurement change as follows:

$$\begin{pmatrix} \text{derivative action on} \\ \text{measurement change} \end{pmatrix} = \begin{pmatrix} \text{rate of change} \\ \text{of measurement} \end{pmatrix} \times \begin{pmatrix} \text{derivative} \\ \text{time} \end{pmatrix} \times \begin{pmatrix} \text{derivative} \\ \text{gain} \end{pmatrix}$$

Use the above information to perform the following tasks.

1. Calculate the derivative action for a modern controller on the measurement value at the instant before and the instant after the changes in the measurement value occur at 0 min, 10 min, and 25 min.

_____

2. Graph the derivative-adjusted measurement value for a modern controller from 0 min to 30 min.

_____

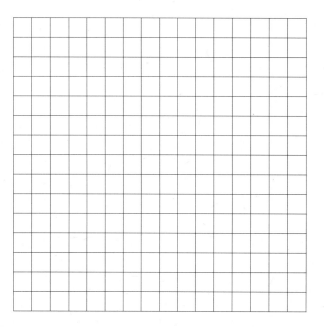

3. Calculate the proportional output for a modern controller at the instant before and the instant after the changes in the measurement value occur at 0 min, 10 min, and 25 min.

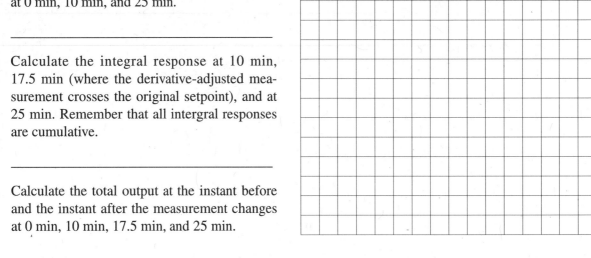

_____

4. Calculate the integral response at 10 min, 17.5 min (where the derivative-adjusted measurement crosses the original setpoint), and at 25 min. Remember that all intergral responses are cumulative.

_____

5. Calculate the total output at the instant before and the instant after the measurement changes at 0 min, 10 min, 17.5 min, and 25 min.

_____

6. Graph the total output for a modern controller from 0 min to 25 min.

SECTION
9 AUTOMATIC CONTROL

chapter

37

Controller Tuning

REVIEW
QUESTIONS

Name _____ **Date** _____

_____ 1. Controller tuning is the process of determining the tuning ___ to obtain a desired controller response to process disturbances.

_____ 2. Common controller performance standards are decay ratio, overshoot, and ___.

_____ 3. The ___ is a measure of how quickly an overshoot decays from one oscillation to the next as the controller brings the process to the setpoint.
    A. decay ratio
    B. undershoot
    C. gain
    D. proportional band

T   F 4. A one-quarter decay ratio is a response where the amount of undershoot decays to one-fourth of the previous amplitude of the overshoot every whole cycle after being upset by a disturbance.

T   F 5. Processes with a lot of dead time must be tuned to respond slowly and minimize overshoot.

_____ 6. ___ is the length of time required for the PV to cross the ultimate value after a step input change, such as a setpoint change.

T   F 7. Static response time is the length of time required for the PV to remain within 5% of its ultimate value following a step input change.

_____ 8. ___ is the factor by which the controller gain may be increased before instability occurs, and therefore is a measure of relative stability.
    A. Throttling range
    B. Adaptive gain
    C. Gain margin
    D. Open loop tuning

T   F 9. A self-tuning controller is a controller that has built-in algorithms or pattern recognition techniques that periodically test the process and make changes to the controller tuning settings while the process is operating.

T   F 10. A stand-alone controller is a controller that has its power supplies, input signal processing, controller functions, output signals, and displays contained in the same case.

_____ 11. ___ is a check that must be made to a control loop prior to placing it in operation.
    A. Ensuring that the controller action is selected correctly
    B. Determining which controller functions are needed for the process
    C. Determining the initial tuning values for the chosen controller functions
    D. all of the above

_____ **12.** The ___ method is a manual tuning method that consists of making small changes in the setpoint and observing the responses.
- A. self-tuning
- B. setpoint step change
- C. Ziegler-Nichols open loop
- D. Ziegler-Nichols closed loop

T　　F　**13.** The tuning map method is a procedure for controller tuning that compares the process curves to one of numerous typical closed loop response curves.

T　　F　**14.** The ultimate period method is also known as the Ziegler-Nichols closed loop tuning method.

_____ **15.** Ziegler-Nichols methods are empirical and intended to achieve a ___.
- A. one-quarter decay ratio
- B. 10% rise time
- C. 10% dynamic response time
- D. 10% static response time

_____ **16.** Ziegler-Nichols ___ tuning is a method of tuning a controller by increasing the gain until the system cycles at the point of instability.
- A. closed loop
- B. open loop
- C. dynamic gain
- D. static gain

_____ **17.** The ___ gain (K_u) is the proportional gain at the point of oscillation.
- A. proportional
- B. integral
- C. minimum
- D. ultimate

T　　F　**18.** The ultimate period (T_u) is the cycle time at the point of oscillation.

_____ **19.** ___ open loop tuning is a method of tuning a controller based on open loop response to a step input.
- A. Ziegler-Nichols
- B. Rise time
- C. Dynamic response
- D. Stand-alone

Name _____   Date _____

Activity 37-1—One-Quarter Decay Ratio

One of several common criteria for controller tuning is the one-quarter decay ratio. Plot the answers on the graphs.

1. Draw a typical one-quarter decay ratio closed response for a load change of a proportional-only controller.

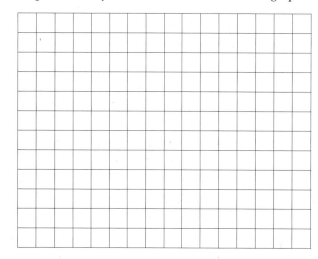

2. Draw a typical one-quarter decay ratio closed response for a load change of a proportional-integral controller.

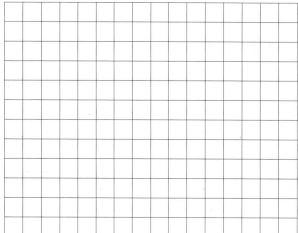

3. List two of the important features of one-quarter decay ratio tuning.

Name _____   **Date** _____

Activity 37-2—Controller Tuning

The following closed loop responses are for a proportional-only controller and a proportional-integral controller. For each curve, state the needed change in tuning values (increase, no change, or decrease).

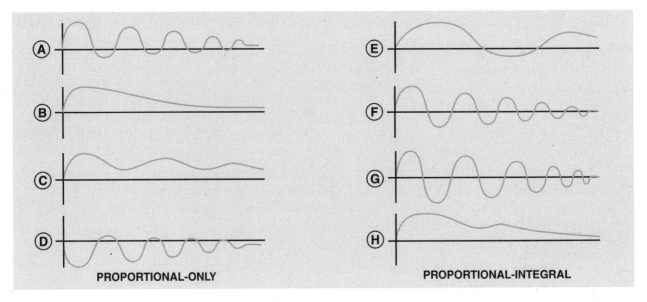

1. Curve A proportional-only.

   _____

2. Curve B proportional-only.

   _____

3. Curve C proportional-only.

   _____

4. Curve D proportional-only.

   _____

5. Curve E proportional-integral.

   _____

6. Curve F proportional-integral.

   _____

7. Curve G proportional-integral.

   _____

8. Curve H proportional-integral.

   _____

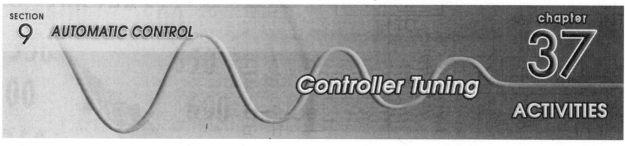

Name _____   **Date** _____

Activity 37-3—Ziegler-Nichols Closed Loop Tuning

Using the Ziegler-Nichols closed loop tuning method, it was found that the ultimate gain setting, K_u, was 1.15 and the ultimate period, T_u, was 2.73 min. Calculate the recommended controller settings for a proportional-integral-derivative controller.

_____ **1.** What is the gain?

_____ **2.** What is the proportional band?

_____ **3.** What is the integral time?

_____ **4.** What is the derivative time?

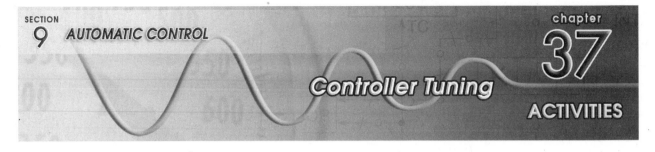

Name _____   **Date** _____

Activity 37-4—Ziegler-Nichols Open Loop Tuning

An open loop response test is conducted on a process. Using the Ziegler-Nichols open loop tuning procedure and equations, calculate the recommended tuning values for a proportional-integral controller.

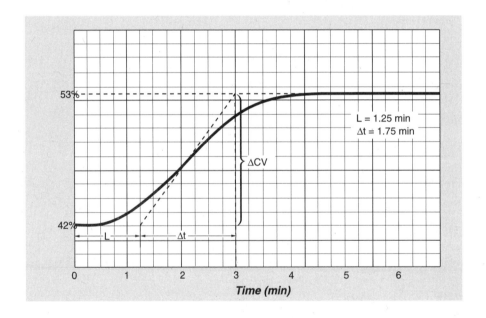

_____   **1.** What is the recommended gain?

_____   **2.** What is the recommended integral time?

Name _____ **Date** _____

T   F   **1.** A pneumatic controller uses microprocessor technology and special programming to perform controller functions.

_____   **2.** ___ is the selection of preprogrammed software packages embedded in a controller representing available features that can be chosen.

_____   **3.** ___ is the time that it takes to access all I/O points and go through a configuration program.

T   F   **4.** A digital control chart is a type of configuration format consisting of a series of conditional statements, parallel paths, and action blocks that begins with a Start command and ends with an existing scan.

_____   **5.** A(n) ___ controller is an electric controller that has DC electric current as the output.

_____   **6.** A(n) ___ is a view into a digital control system through which an operator can observe and control a process.
     A. controller
     B. pressure gauge
     C. operator interface
     D. instrument

T   F   **7.** A stand-alone display system is designed for use with larger digital control systems and has the electronics and display packaged separately in different enclosures since the control system is too large to be packaged with the display.

_____   **8.** A(n) ___ is a display system designed so that touching specific spots on a screen produces an action.

T   F   **9.** A PC-based display system is a display system that uses standard personal computer hardware and special software to display full graphics, control functions, alarms, trends, etc.

_____   **10.** A digital recorder is a device that uses local memory to record process data, which is then displayed on digital ___ that can have the appearance of old-style strip chart recorders.
     A. gauges
     B. display screens
     C. sensors
     D. indicators

_____ **11.** A ___ is one of the various programming methods that have been developed to provide a simplified method for instructing various digital control systems on how to control a process.
    A. programmable logic controller
    B. distributed control system
    C. pneumatic control system
    D. configuration format

_____ **12.** ___ is a configuration format where the configuration is selected from a list of available functions.
    A. Ladder logic
    B. Pick and choose
    C. Function block configuration
    D. Structured programming

T   F   **13.** Function block configuration is a configuration method that uses a library of functions provided by the manufacturer.

T   F   **14.** Ladder logic is a configuration method that consists of two vertical rails, the right one being the source and the left one being the end, and the sequential rungs of logic between the two.

_____ **15.** ___ text is a type of configuration that is very similar to Microsoft® Visual Basic® or older structured programming languages.

Name _____ **Date** _____

Activity 38-1—PLC Troubleshooting

Operating and fault status indicators include the power status, PC run, CPU fault, forced I/O, and battery-low indicators. The power status indicator turns ON to indicate that the processor is energized and power is being applied. This status indicator should normally be ON. The PC run indicator turns ON when the processor is in the run mode. Care must be taken when the run indicator is ON because the controller activates the loads as programmed. This indicator is OFF when the processor is placed in the program mode.

The CPU fault indicator turns ON when the processor has detected an error in the controller. The processor automatically shuts OFF all loads and stops operation when this indicator is ON. The forced I/O indicator turns ON when one or more input or output devices have been forced ON or OFF. All force commands must be removed from the program before normal operation is resumed. A battery is used to provide backup power for the processor memory in case of an external power failure. The battery-low indicator turns ON when the battery should be replaced or if the battery is not charging.

The input/output (I/O) status indicator shows the status of the input and output signals. The status indicators on the input module are energized when an electrical signal is received at an input terminal. This occurs when an input contact is closed or signal is present. The status indicators in the output module are energized when an output signal is sent through that output wire. Each input device and output device has its own status indicator.

Use the Pump Unloader Circuit Line Diagram and the Pump Unloader Circuit Input and Output Connections Diagram on the next page to answer the following questions.

1. The circuit has normal status indication when the PLC is in the run mode. The selector switch is in the run position and the STOP pushbutton has been pressed and released. Fill in the blanks to state whether each indicator light is ON or OFF.

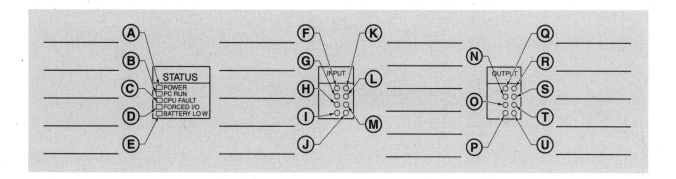

2. The circuit has normal status indication when the PLC is in the run mode. The selector switch is in the run position and the START pushbutton has been pressed and released. Fill in the blanks to state whether each indicator light is ON or OFF.

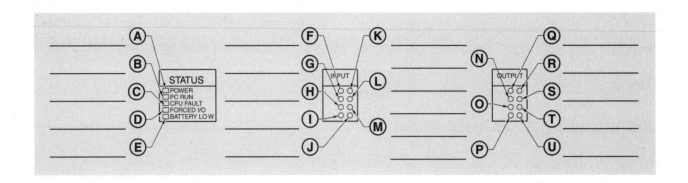

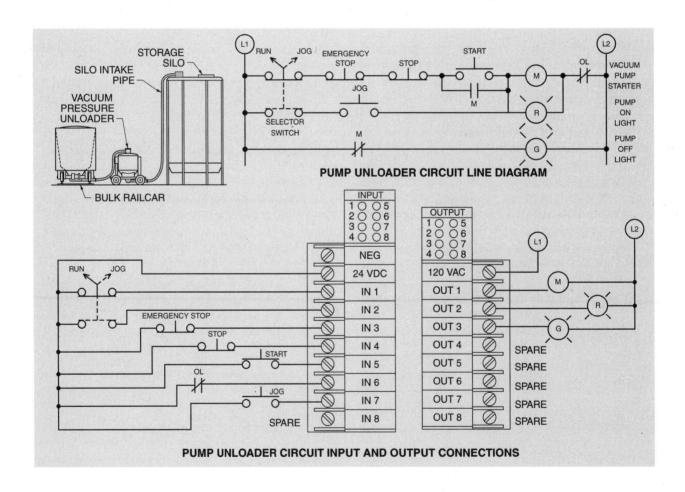

PUMP UNLOADER CIRCUIT LINE DIAGRAM

PUMP UNLOADER CIRCUIT INPUT AND OUTPUT CONNECTIONS

Name _____   **Date** _____

Programming Controllers

System Description

A solar heating system converts solar radiation into heat that may be used to maintain a set temperature in a building. Drawings for this system are on the following pages. A solar collector heats a liquid. Pump 1 circulates the liquid through a heat exchanger/storage tank. Pump 1 automatically energizes when the temperature in the solar collector is 10°F warmer (T1 – T2) than the temperature in the heat exchanger/storage tank. Heated liquid from the solar collector automatically diverts to a purge coil when the temperature in the heat exchanger/storage tank exceeds an upper temperature limit. The purge coil transfers the heat to the outside air.

The furnace blower motor and Pump 2 turn ON when Temperature Switch 4 (T4) calls for heat. Pump 2 circulates heated liquid from the heat exchanger/storage tank over the heat exchange coil located at the furnace intake. Warm air is circulated out of the furnace as cold air moves through the heated coil. An auxiliary heat source automatically supplements the solar heating system if the amount of heat produced from the heat exchange coil is not enough to heat the building.

System Operation

Placing the selector switch in the ON position turns ON the solar heating system. The temperature differential relay contacts close if the temperature differential relay measures a 10°F temperature difference. Pump 1 and the directional control valve solenoid energize when the temperature differential relay contacts close. Pump 1 moves the fluid through the solar collector. The directional control valve directs the heated fluid to the heat exchanger/storage tank.

The purge relay (CR1) energizes if T3 detects excessive heat. The purge relay de-energizes the directional control valve solenoid. This directs the heated fluid through the purge coil. The furnace blower motor, Pump 2, and the 5 min blower timer energize when T4 calls for heat. The auxiliary heat contactor energizes if T4 does not open after 5 min.

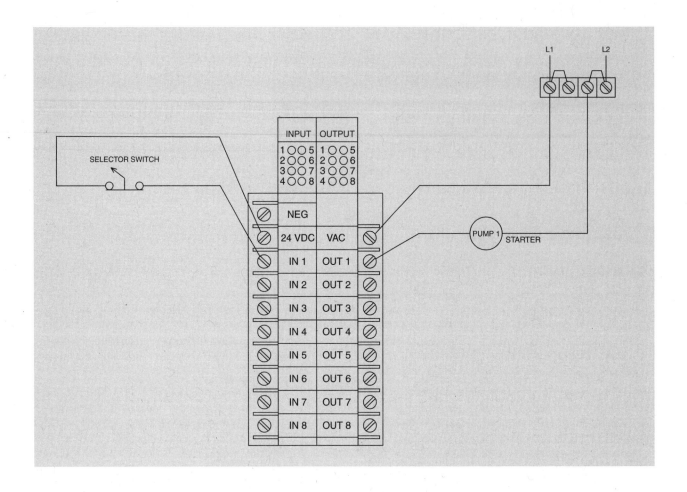

Use the system drawings on the next several pages to perform the following tasks.

1. Identify the seven inputs in the Solar Heating System – Line Diagram. Add the inputs to the drawing. Connect each input to the input module so that the first input in the line diagram is connected to IN 1, and so on. The first input has been added as an example.

2. Identify the five outputs in the Solar Heating System – Line Diagram. Add the outputs to the drawing above. Connect each output to the output module so that the first output in the line diagram is connected to OUT 1, and so on. The first output has been added as an example.

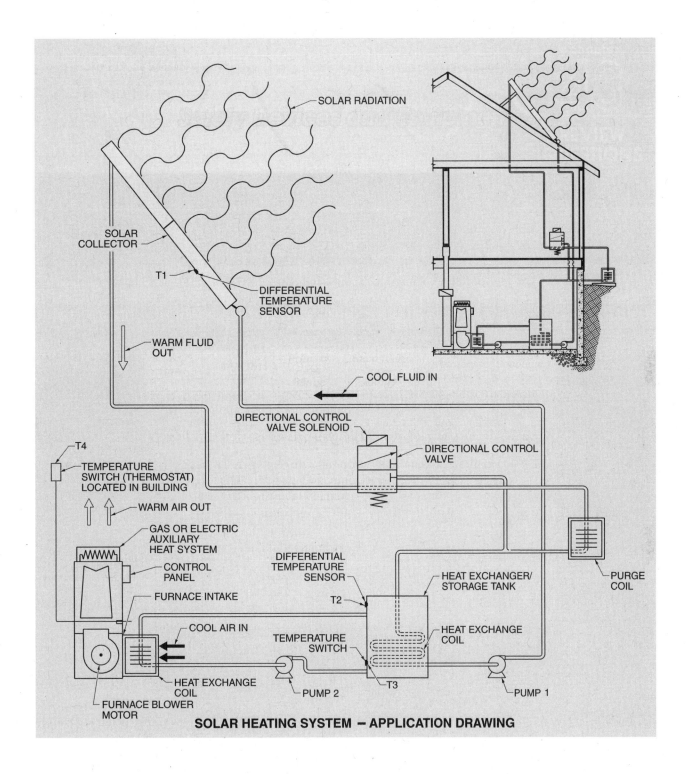

SOLAR HEATING SYSTEM – APPLICATION DRAWING

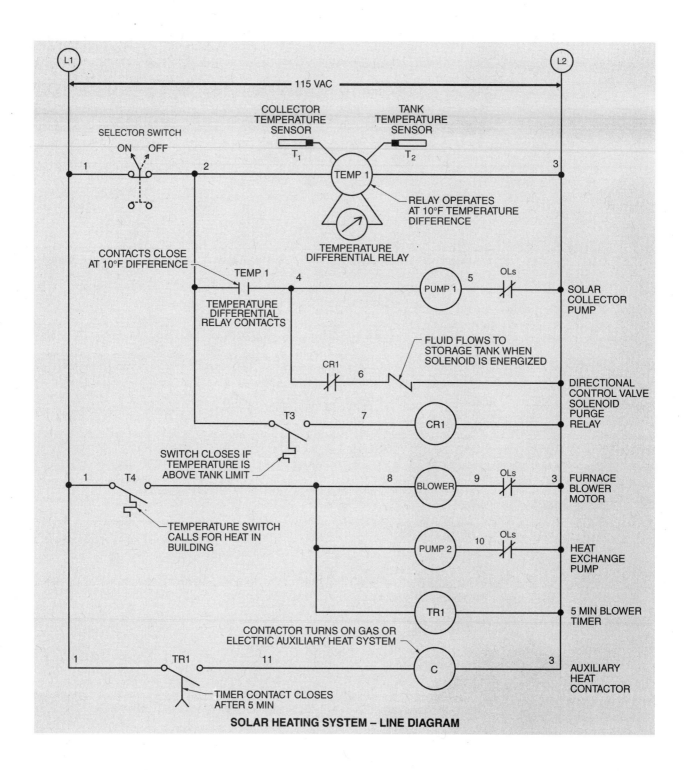

SOLAR HEATING SYSTEM – LINE DIAGRAM

SECTION
10 *FINAL ELEMENTS*

chapter
39

Control Valves

REVIEW
QUESTIONS

Name _____   **Date** _____

_____   **1.** A final element is a device that receives a control signal and ___ the amount of
material or energy in a process.
    A. regulates
    B. increases
    C. decreases
    D. pressurizes

T   F   **2.** A normally open valve does not allow pressurized fluid to flow out of the valve
in the spring-actuated (de-energized) position.

_____   **3.** A(n) ___ valve allows the flow rate percentage to change by an amount equal to
the change in the opening percentage.
    A. equal-percentage
    B. linear
    C. quick-opening
    D. solenoid-actuated

_____   **4.** ___ is the process where a portion of a liquid coverts to a vapor as it passes through
a control valve because the pressure has fallen below the vapor pressure of the
liquid.

_____   **5.** The ___ pressure ratio is the ratio of the downstream pressure to upstream pressure
where the gas velocity out of the valve is a sonic velocity.
    A. sonic
    B. downstream
    C. preferred
    D. critical

_____   **6.** A globe valve is a throttling valve where the flow enters ___, makes a turn through
the plug and seat, and then makes another turn to exit the valve.
    A. axially
    B. vertically
    C. horizontally
    D. circumferentially

T   F   **7.** A three-way mixing globe valve has one inlet and two outlets.

_____ **8.** A valve ___ is a machined disc or shaped piece that regulates the flow by changing the size of the valve opening.
   A. plug
   B. globe
   C. seat ring
   D. stem

_____ **9.** A Class ___ leakage rating for a globe valve means that the leakage is 0.1% or less of valve capacity.
   A. I
   B. II
   C. III
   D. IV

_____ **10.** A ___ valve is a throttling control valve consisting of a one-piece body incorporating an internal weir and a flexible diaphragm.
   A. weir
   B. globe
   C. split body
   D. diaphragm

_____ **11.** A ___ valve is a valve with a disc that is rotated perpendicular to the valve body.
   A. butterfly
   B. globe
   C. sliding stem
   D. ball

_____ **12.** A(n) ___ is a path through a valve from an inlet port to an outlet port.

T   F   **13.** Cavitation causes wear patterns in a valve that resemble a sanded and polished surface.

T   F   **14.** Cavitation is the process where vapor bubbles in a flowing liquid collapse inside a control valve.

T   F   **15.** If the critical pressure ratio for air flowing through a valve is 0.527 and the actual pressure ratio is 0.477, then a choked flow condition exists.

_____ **16.** A ___ globe valve consists of two plugs and seat rings through which the fluid flows.
   A. single-port
   B. double-port
   C. dual-flow
   D. double-dip

T   F   **17.** A ball valve is a throttling valve consisting of a straight-through valve body enclosing a ball with a hole through the center.

_____ **18.** The ___ has established standard drawing symbols for solenoid valves.
   A. Hydraulic Power Institute (HPI)
   B. Joint Industry Conference (JIC)
   C. Pneumatic Standards Team (PST)
   D. Solenoid Manufacturers Association (SMA)

Name _____   Date _____

Activity 39-1—Flow Characteristics

The valve typical characteristic flow curves for three control valves are shown. They are linear, equal-percentage, and quick-opening. Use the percent flow at the 5% and the 100% rated travel positions to answer the following questions.

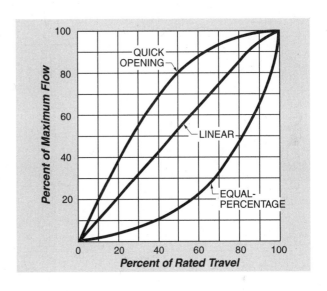

_____   **1.** What is the rangeability for the linear characteristic curve?

_____   **2.** What is the rangeability for the equal-percentage characteristic curve?

_____   **3.** What is the rangeability for the quick-opening characteristic curve?

4. Construct a table of the flow percent vs. the travel percent for the equal-percentage characteristic curve. Read the flow percent at each 10% increase in travel percent from 0% to 100%. Calculate the percent increase in flow for each 10% increase in travel percent.

5. Explain why the equal-percentage characteristic curve is given that name.

_____

_____

_____

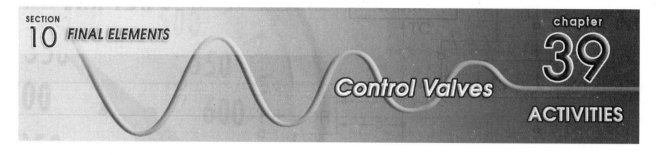

Name _____   Date _____

Activity 39-2—Valve Selection

A control valve is installed in a long 3″ pipeline where, at a design flow of 100 gpm, the pipeline $\Delta p = 43$ psig and the control valve $\Delta p = 7$ psig. The calculated control valve size is $C_v = 39$ at design conditions. At minimum flow of 10 gpm, the calculated C_v is 1.5.

				Linear Single Port (ES)										Linear Characteristic
Coeffi-cients	Body Size, In.	Port Diameter, In.	Total Travel, In.	Valve Opening—Percent of Total Travel										K_m and C_1
				10	20	30	40	50	60	70	80	90	100	
C_v (Liquid)	1 & 1¼	1⁵⁄₁₆	¾	2.27	4.12	6.23	8.54	11.0	13.4	15.8	17.8	19.3	20.1	0.80
	1½	1⅞	¾	3.56	7.01	11.1	15.1	19.0	22.9	26.7	30.0	33.1	34.9	0.85
	2	2⁵⁄₁₆	1⅛	8.49	17.1	25.9	35.3	44.4	52.9	59.2	62.0	63.9	65.3	0.83
	2½	2⅞	1½	10.4	22.2	34.9	47.1	58.2	66.6	73.7	79.3	84.4	86.5	0.86
	3	3⁷⁄₁₆	1½	15.3	34.3	52.8	71.4	87.8	101	112	121	129	135	0.80
	4	4⅜	2	23.7	46.4	72.9	98.2	122	145	165	183	199	212	0.79
	6	7	2	55.0	118	180	135	280	312	341	368	390	417	0.66
	8	8	2	66.6	147	221	292	375	450	522	592	652	701	0.70
	8	8	3	100	213	330	451	553	648	719	773	809	836	0.72

				Equal-Percentage Single Port (ES)										Equal-Percentage Characteristic
Coeffi-cients	Body Size, In.	Port Diameter, In.	Total Travel, In.	Valve Opening—Percent of Total Travel										K_m and C_1
				10	20	30	40	50	60	70	80	90	100	
C_v (Liquid)	1 & 1¼	1⁵⁄₁₆	¾	0.783	1.29	1.86	2.71	4.18	6.44	9.54	13.1	15.7	17.4	0.90
	1½	1⅞	¾	1.54	2.52	3.57	4.94	7.41	11.6	17.2	23.5	28.7	33.4	0.89
	2	2⁵⁄₁₆	1⅛	1.74	3.15	4.72	6.91	10.6	16.3	25.0	36.7	47.8	56.2	0.85
	2½	2⅞	1½	4.05	7.19	10.6	14.5	21.2	31.6	45.5	64.2	77.7	82.7	0.87
	3	3⁷⁄₁₆	1½	4.05	6.84	10.0	15.0	23.8	37.8	59.0	87.1	110	121	0.79
	4	4⅜	2	6.56	11.4	17.3	27.0	42.2	66.4	103	146	184	203	0.82
	6	7	2	13.2	24.6	41.1	62.5	97.1	155	223	286	326	357	0.74
	8	8	2	18.8	33.6	53.6	79.8	114	168	242	345	467	570	0.72
	8	8	3	25.9	53.3	97.8	178	299	461	618	727	768	808	0.72

1. From the catalog sizing information, select the proper size valve. Give the valve size and tell whether it has linear or equal-percentage characteristics.

_____

_____

Name _____   **Date** _____

Activity 39-3—Valve Sizing

Hot condensate at 220°F from a collection tank is pumped up 20 ft through a control valve into a deaerator. The flow out of the pump is 15 gpm at 25 psig and 220°F. The control valve at the entrance to the deaerator needs to be sized to handle the flow. There is always the possibility of flashing when handling hot condensate. Flashing limits the Δp to the difference between the fluid vapor pressure and the inlet pressure.

Use the flow equation, $F = C_v \sqrt{\Delta p}$, to answer the following question.

1. What is the required C_v for the control valve?

   _____

2. Select the best valve from the valve sizing table.

   _____

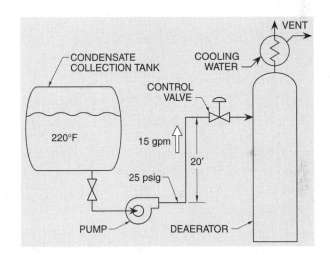

	Equal-Percentage Single Port (ES)												Equal-Percentage Characteristic
Coeffi-cients	Body Size, In.	Port Diameter, In.	Total Travel, In.	Valve Opening—Percent of Total Travel										K_m and C_1
				10	20	30	40	50	60	70	80	90	100	
					Standard Port									
C_v (Liquid)	1 & 1¼	1⁵⁄₁₆	¾	0.783	1.29	1.86	2.71	4.18	6.44	9.54	13.1	15.7	17.4	0.90
	1½	1⅞	¾	1.54	2.52	3.57	4.94	7.41	11.6	17.2	23.5	28.7	33.4	0.89
	2	2⁵⁄₁₆	1⅛	1.74	3.15	4.72	6.91	10.6	16.3	25.0	36.7	47.8	56.2	0.85
	2½	2⅞	1½	4.05	7.19	10.6	14.5	21.2	31.6	45.5	64.2	77.7	82.7	0.87
	3	3⁷⁄₁₆	1½	4.05	6.84	10.0	15.0	23.8	37.8	59.0	87.1	110	121	0.79
					Restricted Port									
	1½	1⁵⁄₁₆	¾	0.882	1.35	1.89	2.52	3.68	5.52	8.13	12.0	16.5	21.0	0.91
	2	1⁵⁄₁₆	¾	0.849	1.34	1.83	2.39	3.43	5.12	7.49	11.2	15.8	20.8	0.82
	2½	1⅞	¾	1.43	2.37	3.34	4.76	7.25	11.3	17.3	24.2	31.8	40.3	0.90
	3	2⁵⁄₁₆	1⅛	2.74	3.44	4.86	6.95	10.6	16.5	25.0	37.7	52.7	67.5	0.88
	4	2⅞	1½	3.96	7.14	10.6	14.5	21.1	31.7	48.0	69.7	95.6	121	0.88

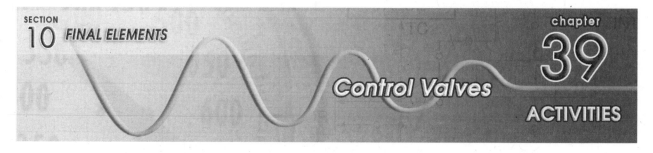

Name _____ Date _____

Activity 39-4—Valve Noise

The primary steam feed to a multistage evaporator uses 250 psig saturated steam. It is fed to the primary steam heat exchanger that transfers heat to the most concentrated solution in the evaporator. Under normal operating conditions, the pressure in the steam chest is 180 psig with a control valve $\Delta p = 70$ psig. During startup conditions with diluted material in the evaporator system, the steam chest pressure is only 90 psig. This places the system into the critical pressure range and it produces a lot of noise, requiring a special valve trim to quiet the valve during startups. The noise at the startup conditions with a standard valve was 83.5 dB. The noise at the startup conditions with a Quiet Trim I valve was 66 dB.

Flow rate (normal) = 57,000 lb/hr; $\Delta p = 70$ psi; $C_s = 280$
Flow rate (startup) = 57,000 lb/hr; $\Delta p = 160$ psi; $C_s = 220$

Single Seat (ES) Globe Flow Coefficients

Coeffi-cients	Body Size, In.	Port Diameter, In.	Max Travel, In.	Quiet Trim I — Valve Opening—Percent of Maximum Travel										Linear Characteristic K_m and C_1
				10	20	30	40	50	60	70	80	90	100	
C_S (Steam)	1 & 1¼	1⁵⁄₁₆	¾	5.75	12.1	17.6	22.8	26.5	28.1	28.7	29.0	29.3	30.3	32.9
	1½	1⅞	¾	5.45	14.2	22.5	30.0	37.5	44.9	51.0	55.5	58.0	59.5	32.0
	2	2⁵⁄₁₆	1⅛	13.3	28.0	41.6	56.0	70.5	85.0	96.0	103	108	110	32.4
	2½	2⅞	1½	24.8	46.5	67.5	88.0	109	126	136	144	148	150	32.7
	3	3⁷⁄₁₆	1½	34.0	58.0	81.0	106	129	152	172	187	198	204	31.1
	4	4⅜	2	55.3	102	141	184	227	265	297	318	332	339	32.1
	6	7	2	69.0	153	222	280	348	412	468	516	556	585	29.2

1. With the information provided, select the best control valve using Quiet Trim I.

   _____

   _____

Name _____   Date _____

Activity 39-5—Sliding Stem and Rotary Valves

It is general practice when selecting a control valve to satisfy an application that the C_v calculated for the design conditions be between 70% to 80% of the travel of an equal-percentage characteristic, about 60% of a linear characteristic, about 45° of a 0° to 60° butterfly valve, and about 60° of a 0° to 90° rotary valve like a V-ball. When a minimum flow condition C_v is provided, the selected valve must cover both design and minimum conditions. Butterfly valves generally cannot throttle at less than 20° open. In some cases, it is suitable to select the design C_v at about the 90% travel position if it is impossible to have more flow than the design conditions. Do not overlook the use of the restricted trim globe valves.

V-Ball Valve

| | | 1.5 to 1 (Line- to Valve-Size Ratio) | | | | | | | Approximately Equal Percentage Characteristic | |
| Coeffi-cient | Valve Size, Inches | Valve Rotation, Degrees | | | | | | | | |
		10	20	30	40	50	60	70	80	90
C_v (Liquid)	3	0.169	7.77	24.1	42.8	71.5	108	146	214	257
	4	0.108	9.2	34.2	66.9	106	154	221	314	427
	6	0.996	20.8	56.9	115	194	285	394	567	886
	8	1.41	29.6	95	197	316	474	684	1032	1480
	10	7.3	74	199	380	604	879	1250	1770	2450
	12 (V150, V300)	7.5	112	291	542	880	1270	1740	2380	3380
	14 (V150)	56.0	232	501	805	1130	1520	2060	2950	4380
	16 (V150)	30.0	237	599	1030	1480	2000	2780	4140	6160
	20 (V150)	105	435	941	1510	2120	2860	3870	5570	8330

Butterfly Valve (Conventional Disc)

| Coeffi-cients | Valve Size, Inches | Disc Angle of Opening, Degrees | | | | | | | | |
		10	20	30	40	50	60	70	80	90
C_v (Liquid)	2	0.2	1.8	5.7	12.7	24.0	40.1	71.4	86.7	91.2
	3	0.5	5.1	16.1	35.8	67.6	112	200	243	256
	4	1.0	10.3	32.6	72.5	136	227	405	492	518
	6	22.7	55.9	131	244	454	769	1120	1610	1750
	8	36.6	90.2	211	394	733	1240	1804	2590	2820
	10	60.2	148	347	648	1200	2040	2960	4260	4630
	12	91.2	224	526	982	1820	3090	4490	6460	7020

Globe Valve (All TFE Construction)

Valve Size In. (DN)	Orifice Size In. (mm)	Range-ability	Minimum Controllable C_v	C_v at 10% Travel Increments									
				10	20	30	40	50	60	70	80	90	100
1" (25)	0.250" (6.35)	25:1	0.004	0.006	0.008	0.011	0.014	0.020	0.028	0.038	0.053	0.072	0.10
			0.006	0.009	0.012	0.017	0.023	0.032	0.044	0.061	0.084	0.12	0.16
			0.010	0.014	0.019	0.026	0.036	0.050	0.069	0.095	0.13	0.18	0.25
			0.016	0.022	0.030	0.042	0.058	0.080	0.11	0.15	0.21	0.29	0.40
			0.025	0.035	0.048	0.066	0.091	0.13	0.17	0.24	0.33	0.46	0.63
	0.562" (14.27)	50:1	0.020	0.030	0.044	0.065	0.096	0.14	0.21	0.31	0.46	0.68	1.00
			0.032	0.047	0.070	0.10	0.15	0.23	0.33	0.49	0.73	1.08	1.60
			0.050	0.07	0.11	0.16	0.24	0.35	0.52	0.77	1.14	1.69	2.50
			0.080	0.12	0.17	0.26	0.38	0.57	0.84	1.24	1.83	2.70	4.00
			0.100	0.15	0.22	0.32	0.48	0.71	1.05	1.55	2.29	3.38	5.00
	0.875" (22.22)	50:1	0.126	0.19	0.28	0.41	0.60	0.89	1.32	1.95	2.88	4.26	6.30
			0.150	0.22	0.33	0.49	0.72	1.06	1.57	2.32	3.43	5.07	7.50
			0.200	0.30	0.44	0.65	0.96	1.41	2.09	3.09	4.57	6.76	10.0
1½" (40)	0.875" (22.22)	50:1	0.126	0.19	0.28	0.41	0.60	0.89	1.32	1.95	2.88	4.26	6.30
			0.150	0.22	0.33	0.49	0.72	1.06	1.57	2.32	3.43	5.07	7.50
	1.500" (38.10)	50:1	0.200	0.30	0.44	0.65	0.96	1.41	2.09	3.09	4.57	6.76	10.0
			0.300	0.44	0.66	0.97	1.43	2.12	3.14	4.64	6.86	10.1	15.0
			0.500	0.62	0.92	1.36	2.01	2.97	4.39	6.49	9.60	14.2	21.0

A flow control loop feeds 31% HCl acid to a steel pickling bath. The flow range is 0 gpm to 15 gpm and the design flow is 10.0 gpm. The design flow C_v = 3.45. The piping size is 1½" TFE lined pipe.

1. Select a TFE globe valve. What size valve and control range should be used?

_____

_____

Cooling water is fed to a heat exchanger through a 6" pipe. The design C_v = 104 and the control valve is only required to have a 4:1 turndown. It is desired to use an inexpensive conventional butterfly valve.

2. What size valve and control range should be selected?

_____

_____

Chilled water is used in an air conditioning system to provide cool air. Summer design conditions call for a valve with a C_v = 970 with a turndown of 50:1. The line size is 12".

3. What type and size valve should be selected?

_____

_____

A plant waste treatment plant requires the use of 15% caustic hydroxide solution for the neutralization of acid wastes. The design case requires a C_v = 0.041.

_____ 4. Select the type and size of the control valve to satisfy this application.

SECTION
10 FINAL ELEMENTS

chapter
40

Regulators and Dampers

REVIEW
QUESTIONS

Name _____   Date _____

      T   F   **1.** Regulators require no other source of energy than the process itself.

_____ **2.** A pressure regulator is an adjustable valve that is designed to automatically control the pressure ___.
    A. downstream of the regulator
    B. upstream of the regulator
    C. inside a pressure vessel
    D. at a control valve

_____ **3.** The three basic components of most regulators are a ___, a primary element, and a final element.
    A. pressure sensor
    B. secondary element
    C. loading mechanism
    D. positioner

_____ **4.** ___ is a drop in pressure below the set value when there is a high flow demand through a regulator.
    A. Eccentric flow
    B. Lockup
    C. Flow fault
    D. Droop

_____ **5.** A(n) ___ pressure regulator consists of a throttling element, such as a valve plug, connected to a pressure-sensing diaphragm that is opposed by a spring.

_____ **6.** A(n) ___ pressure regulator uses air pressure instead of the force of a spring to oppose the downstream pressure.

_____ **7.** A pilot-operated pressure regulator uses the ___ as a pressure source to power the diaphragm of a larger valve.
    A. downstream fluid pressure
    B. upstream fluid pressure
    C. pneumatic relay
    D. controller air pressure signal

_____ **8.** The purpose of a pilot-operated pressure regulator is to provide more accurate pressure control by nearly eliminating ___.
  A. flow measurement errors
  B. droop
  C. calibration errors
  D. temperature compensation

T   F   **9.** A differential pressure regulator controls the pressure difference between the outlet pressure of the regulator and the fluid loading pressure.

_____ **10.** A ___ regulator maintains the pressure upstream of the regulator to a set value.
  A. backpressure
  B. pressure-relief
  C. differential pressure
  D. modulating

_____ **11.** A(n) ___ is an adjustable blade or set of blades used to control the flow of air.

_____ **12.** A(n) ___-blade damper is a damper in which adjacent blades are parallel and move in the same direction with one another.
  A. opposed
  B. adjacent
  C. parallel
  D. round

T   F   **13.** Opposed-blade dampers are more expensive but provide better flow characteristics than parallel-blade dampers.

_____ **14.** A temperature self-operating regulator is a combination of a thermal filled system and a(n) ___.

Name _____   Date _____

Activity 40-1—Regulator Sizing

Use the pressure regulator specification sheets on the following page to answer the questions. In all cases, the design flow is 10% offset.

_____ 1. What is the best regulator to use when a pressure-reducing regulator set at 10 psig and using 100 psig air supply is used to purge a header with 20 scfm of air when an air-operated valve is opened?

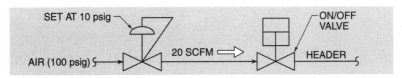

_____ 2. What is the best regulator to use when a pressure-reducing regulator set at 5 psig and using 50 psig nitrogen supply blankets the vessel with 8 scfm when an automatic valve opens?

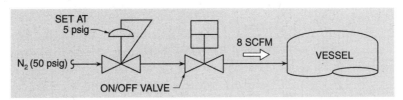

_____ 3. What is the best regulator to use when a differential pressure regulator set at a 10 psig differential sets a steam pressure to follow a fuel oil pressure? At a 20 psig oil pressure, the steam flow is to be 60 lb/hr.

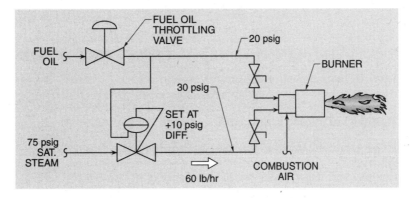

4. What is the best regulator to use when a pressure-reducing regulator is used to supply saturated steam to a heat exchanger with 100 lb/hr steam at 30 psig? The source is 125 psig steam.

Air SCFH Capacities for ¼- through 1-Inch Types 5L and 5LD Regulators with Elastomer Diaphragm

Recommended Outlet Pressure Range	Pressure			Regulator Body Size, Inches							
	Outlet Setting	Inlet		¼		½		¾		1	
		psig	bar	10% Offset	20% Offset	10% Offset	20% Offset	10% Offset	20% Offset	10% Offset	20% Offset
2 psig to 6 psig (0.1 bar to 0.4 bar)	5 psig (0.3 bar)	10	0.7	200	375	400	600	1100	1700	1300	2000
		20	1.4	300	525	500	750	1850	3400	2200	4000
		30	2.0	400	600	550	850	2300	4200	2800	5000
		50	3.5	450	600	600	850	3500	5500	4200	6500
		75	5.2	450	625	650	900	3800	6800	4500	8000
		100	6.9	500	625	700	900	4200	7200	5000	8500
		150	10.3	500	625	700	900	4200	7600	5000	9000
		200	13.8	500	625	700	900	4200	8300	5000	9800
		250	17.2	500	625	700	900	4200	8300	5000	10,000
5 psig to 15 psig (0.3 bar to 1.0 bar)	10 psig (0.7 bar)	20	1.4	500	675	750	1100	1700	2500	2000	3000
		30	2.0	600	850	900	1200	2100	4200	2500	5000
		50	3.5	750	926	1000	1350	2800	5300	3300	6300
		75	5.2	750	926	1000	1350	4700	8500	5500	10,000
		100	6.9	750	1000	1000	1400	6400	9300	7500	11,000
		150	10.3	750	1000	1000	1400	7700	12,000	9000	14,000
		200	13.8	750	1000	1000	1400	8500	13,000	10,000	15,000
		250	17.2	750	1000	1000	1400	8500	13,000	10,000	15,000

Steam Capacities, in lb/hr, for ¼- through 1-Inch Types 5H and 5HD Regulators with Stainless Steel Diaphragms

Recommended Outlet Pressure Range	Pressure			Regulator Body Size, Inches							
	Outlet Setting	Inlet		¼		½		¾		1	
		psig	bar	10% Offset	20% Offset	10% Offset	20% Offset	10% Offset	20% Offset	10% Offset	20% Offset
15 psig to 30 psig (1.0 bar to 2.0 bar)	15 psig (1.0 bar)	30	2.0	8	12	22	33	37	72	44	85
		40	2.8	10	14	24	40	50	89	58	105
		50	3.5	11	16	27	45	58	100	68	120
		75	5.2	13	20	32	54	85	160	100	190
		100	6.9	15	24	39	58	100	180	120	210
		150	10.3	20	32	45	62	140	240	160	280
		200	13.8	25	41	45	62	170	280	200	330
		250	17.2	26	44	45	62	200	360	240	430
	30 psig (2.0 bar)	40	2.8	11	21	30	53	50	110	60	130
		50	3.5	13	24	36	62	68	140	80	160
		75	5.2	18	33	45	80	90	180	110	210
		100	6.9	23	40	52	86	120	210	140	260
		150	10.3	32	54	62	95	160	270	190	320
		200	13.8	32	60	73	100	190	410	220	490
		250	17.2	35	70	75	100	200	430	240	510

SECTION
10 FINAL ELEMENTS

chapter
41

Actuators and Positioners

REVIEW
QUESTIONS

Name _____   Date _____

_____ **1.** A(n) ___ is a device that provides the power and motion to manipulate the moving parts of a valve or damper used to control fluid flow.
   A. actuator
   B. pilot-operated regulator
   C. ratio regulator
   D. sliding stem

_____ **2.** A(n) ___ actuator consists of a large-diameter diaphragm chamber with a diaphragm backed by a plate attached to the actuator stem and opposed by the actuator spring.
   A. internal control
   B. pneumatic
   C. direct-acting
   D. diaphragm-and-spring

T   F   **3.** Because a pneumatic sliding stem piston actuator has no spring, it normally has no fail-safe position.

_____ **4.** A ___ transducer is a device that can convert an electronic controller output signal into a standard pneumatic output.
   A. pneumatic-to-current (P/I)
   B. voltage-to-pneumatic (V/P)
   C. current-to-pneumatic (I/P)
   D. pneumatic-to-voltage (P/V)

_____ **5.** A(n) ___ is a device used to ensure positive position of a valve or damper actuator.
   A. calibrator
   B. positioner
   C. actuator
   D. characteristic cam

_____ **6.** ___ operation is a control configuration where a single control signal is directed to two or more control valves.
   A. Control-dividing
   B. Alternative valve
   C. Dual-choice
   D. Split range

_____ **7.** Common requirements for an actuator include ___.
  A. speed
  B. power
  C. precision
  D. all of the above

T F **8.** Actuators designed for rotary valves and for dampers typically have shorter strokes than actuators for sliding stem valves.

_____ **9.** A diaphragm-and-spring actuator converts an air pressure change at the diaphragm into a ___.
  A. mechanical movement
  B. voltage
  C. current
  D. pneumatic signal

_____ **10.** The amount of movement of the diaphragm in a diaphragm-and-spring actuator is determined by the ___.

T F **11.** In a reverse-acting actuator, the valve plug moves toward the seat when air is applied.

T F **12.** An advantage of a pneumatic sliding stem piston actuator is that higher air pressures can be used to provide greater power to a valve.

_____ **13.** A three-way solenoid valve is a solenoid that shuts off the air supply and ___ when the solenoid is de-energized.
  A. reverses the process fluid flow
  B. activates an alarm
  C. pressurizes gas in the actuator
  D. vents air from the actuator

_____ **14.** A ___ -way solenoid valve is a single- or dual-coil solenoid that has an air supply, a vent, and two cylinder ports.
  A. one
  B. two
  C. three
  D. four

_____ **15.** A spring-return actuator is a piston or diaphragm actuator with a(n) ___ to force the actuator shaft to one end of its travel.
  A. cam
  B. solenoid
  C. air connection
  D. internal spring

Name _____ Date _____

Activity 41-1—Throttling Actuators

The selection of actuators is a complex task best left to the manufacturer. However, it is helpful to understand the factors affecting the selection process. The force required to operate a control valve must include the following:

A. Force to overcome static unbalance of the valve plug

B. Force to provide seat load

C. Force to overcome packing friction

D. Additional forces required (only included for certain specific applications or constructions)

Total force required = A + B + C + D

A. Force to overcome static unbalance
  Unbalanced valve plug, flow up (push down to close)
  $Force\ A = (P_1 - P_2) \times A_{PORT} + P_2 \times A_{STEM}$
  where
  P_1 = upstream pressure, in psig
  P_2 = downstream pressure, in psig
  A_{PORT} = area of port, in sq in.
  A_{STEM} = area of stem, in sq in.

B. Force to provide seat load
  Class II leakage
  Force B = 20 lb per linear inch of port circumference

C. Stem friction

D. Additional forces
  Only when specified

Valve Detail

Port Diameter	Port Circumference	Port Area	Stem Diameter	Stem Area	Stem Friction	Yoke Boss
1⁵⁄₁₆″	4.12″	1.35	³⁄₈″	0.11	38 lb	2⅛″
2⁵⁄₁₆″	7.26″	4.20	½″	0.20	50 lb	2¹³⁄₁₆″
4⅜″	13.74″	15.03	½″	0.20	50 lb	2¹³⁄₁₆″

Actuator Detail

Actuator Type	VSC, Inches	Yoke Boss Size, Inches	Actuator Size	Effective Diaphragm Area, Inches²
13	⁵⁄₁₆, ³⁄₈	1¼	20	26
13R	³⁄₈	2⅛	32	26
	³⁄₈	2⅛	30	46
			34	69
57 57R 67	½	2¹³⁄₁₆	40	69
			45	105
			46	156
	¾	3⁹⁄₁₆	50	105
			60	156
			70	220

236 INSTRUMENTATION WORKBOOK

The valves selected to have actuators sized for them are all fail closed (air to open). The plant wants to limit the choices for actuators in order to limit the spare parts inventory that is kept on hand. Therefore, the Type 67 diaphragm-and-spring actuator should be specified if possible. The first step is to match the actuator yoke boss size to the valve yoke boss, then look for an actuator that provides the required force.

Type 67—Size 40 (2¹³/₁₆ Inch Yoke Boss)

3-15 psig to Diaphragm

Travel, in.	3/8	7/16	1/2	5/8	3/4	7/8	1 1/8	1 1/2	Spring Part Number (Spring Rate, lb/in.)
	207 (3-15)								52
	345 (5-15)	207 (3-15)							51
	483 (7-15)	414 (6-15)	276 (4-15)						55
	552 (8-15)	483 (7-15)	345 (5-15)	207 (3-15)					49
	552 (8-15)	483 (7-15)	414 (6-15)	276 (4-15)					54
Force, lb (Bench Set, psig)	621 (9-15)	552 (8-15)	483 (7-15)	345 (5-15)	207 (3-15)				58
	690 (10-15)	621 (9-15)	552 (8-15)	414 (6-15)	345 (5-15)	207 (3-15)			57
	759 (11-15)	690 (10-15)	690 (10-15)	552 (8-15)	483 (7-15)	414 (6-15)	207 (3-15)		53
		759 (11-15)	690 (10-15)	621 (9-15)	552 (8-15)	414 (6-15)	207 (3-15)		56
			759 (11-15)	690 (10-15)	721 (9-15)	552 (8-15)	345 (5-15)		72
				759 (11-15)	690 (10-15)	621 (9-15)	483 (7-15)		71

6-30 psig to Diaphragm

Travel, in.	3/8	7/16	1/2	5/8	3/4	7/8	1 1/8	1 1/2	Spring Part Number (Spring Rate, lb/in.)
	621 (9-30)	414 (6-30)							50
	897 (13-30)	690 (10-30)	483 (7-30)						47
	1104 (16-30)	897 (13-30)	759 (11-30)	414 (6-30)					48
	1242 (18-30)	1104 (16-30)	966 (14-40)	690 (10-30)	414 (6-30)				52
Force, lb (Bench Set, psig)	1380 (20-30)	1242 (18-30)	1173 (17-30)	966 (14-40)	690 (10-30)	483 (7-30)			51
	1518 (22-30)	1449 (21-30)	1311 (19-30)	1173 (17-30)	966 (14-40)	759 (11-30)	414 (6-30)		55
		1518 (22-30)	1380 (20-30)	1242 (18-30)	1104 (16-30)	897 (13-30)	552 (8-30)		49
		1518 (22-30)	1449 (21-30)	1311 (19-30)	1173 (17-30)	966 (14-40)	690 (10-30)		54
			1518 (22-30)	1380 (20-30)	1242 (18-30)	1104 (16-30)	828 (12-30)	414 (6-30)	58

A valve has the following characteristics: flow control valve, fail closed action, 2″ body, 2⁵⁄₁₆″ port diameter, 1⅛″ travel, and 2¹³⁄₁₆″ yoke boss. During normal flow conditions, the upstream pressure P_1 = 40 psig, downstream pressure P_2 = 33 psig, and Δp = 7 psi. During conditions of maximum flow, the upstream pressure is 50 psig and the maximum Δp = 50 psig.

_____ 1. What is the force required to overcome the static unbalance during normal operating conditions?

_____ 2. What is the force required to overcome the static unbalance during maximum flow conditions?

3. Select the actuator size and air loading range.

_____

_____

A valve has the following characteristics: level control valve, fail closed action, 2″ body, 2⁵⁄₁₆″ port diameter, ¾″ travel, 2¹³⁄₁₆″ yoke boss, P_1 = 16.34 psig, P_2 = 0 psig, Δp = 16.34 psi, and maximum Δp = 16.34 psi.

4. Select the actuator size and air loading range.

_____

_____

Actuators and Positioners

Name _____   Date _____

Activity 41-2—Solenoid Sizing

A low-pressure nitrogen gas purge system is designed to turn on when power is shut off. The nitrogen pressure is 12 psig and the required C_V is 1.9. The power is 115 VAC.

Pipe Size (in.)	Orifice Size (in.)	C_V Flow Factor	Operating Pressure Differential (psi)					Max. Fluid Temp. °F		Brass Body		S.S. Body	
			Min.	Max. AC		Max. DC		AC	DC	Catalog Number	Constr. Ref. No.	Catalog Number	Constr. Ref. No.
				Air Inert Gas	Water	Air Inert Gas	Water						
NORMALLY CLOSED (Closed when de-energized)													
⅜	⅜	1.8	0	7	5	3	3	180	120	G10	1	G64	1
⅜	⅜	1.8	0	15	15	3.5	3.5	180	150	G13	2	G65	2
½	⁷⁄₁₆	2.8	0	4	6			180		G16	3	G66	3
½	⁷⁄₁₆	2.8	0	15	15	6	6	200	180	A17	4	A67	4
¾	¾	5	0	2	2	1	1	180	150	G3	9		
¾	¾	5	0	4	4			180		G43	9		
¾	⅝	5.4	0	2.5	2.5			180				G63	10
NORMALLY OPEN (Open when de-energized)													
⅜	⅜	1.6	0	15	15	2.4	2	200	180	A70	5		
½	⁷⁄₁₆	2.2	0	15	15	2.5	2	200	180	A71	6		
½	¾	5	0	2	2			180		G82	7		
¾	¾	5.5	0	2	2			180		G83	8		

Select the proper two-way direct-acting solenoid from the catalog information.

_____ **1.** What is the body size?

_____ **2.** What is the orifice size?

_____ **3.** What is the pressure limit?

_____ **4.** What is the material of construction?

_____ **5.** What is the catalog number?

Name _____   Date _____

Activity 41-3—ON/OFF Actuators

ON/OFF valves are usually standard process quarter-turn valves such as plug valves, ball valves, and butterfly valves, with nonthrottling piston-type actuators. The valves are usually pipeline size. The actuator size and operating pressure are selected to overcome the breakaway resistance of the valve.

Valve Operating Torques (in inch-pounds)					
Sleeved Plug Valves		Lined Plug Valves		Lined Ball Valves	
Size	Break Torque	Size	Break Torque	Size	Break Torque
½	140	½	260	1	115
¾	140	¾	260	1½	165
1	400	1	400	2	295
1½	800	1½	600	3	720
2	1100	2	800	4	1210
3	1200	3	1200	6	2520
4	2400	4	1800		
6	5000	6	4800		
8	7800	8	15,000		
10	14,400				
12	21,000				

The two common forms of piston-type actuators are spring-return and double-acting actuators. A 90° rotation, vane-type actuator is double-acting. However, it can be supplied with an air volume tank and special trip valve to make the vane fail to the correct position. This is rarely used. Actuator torques for different air pressures and sizes are shown below. The recommended safety factor is 10% for these actuators.

Actuator Torques (in inch-pounds)				
Piston-Type Double-Acting				
Model	40 psi	60 psi	80 psi	100 psi
D 12	59	89	119	149
D 25	106	168	221	274
D 40	204	310	416	513
D 65	310	149	628	788
D 100	460	499	929	1168
D 200	1009	1522	2044	2558
D 350	1761	2655	3549	4443
D 500	2339	3531	4722	5913
D 750	3479	5251	7022	8794
D 1100	5131	7744	10,357	12,970
D 2500	11,825	17,847	23,869	29,891
D 4000	19,962	30,127	40,293	50,458

Model	Spring Set Number	40 psi		60 psi		80 psi		100 psi	
		Air Start	Air End	Air Start	Air End	Air Start	Air End	Air Start	Air End
	2	196	117	364	285	532	453	700	621
	3	429	10	294	178	466	347	634	515
S65	4			230	71	398	240	567	408
	5					331	133	499	301
	6							432	194
	2	303	192	552	440	800	689	1049	937
	3	211	44	460	292	709	541	957	790
S100	4			367	143	616	373	865	642
	5					523	245	772	494
	6					430	96	679	345
	2	656	406	1201	952	1747	1498	2293	2044
	3	448	74	994	620	1539	1165	2085	1710
S200	4			786	287	1331	832	1878	1378
	5					1124	500	1670	1045
	6					916	167	1462	713
	2	1105	684	2053	1632	3002	2582	3950	3529
	3	727	95	1675	1043	2623	1991	3571	2939
S350	4			1296	454	2244	1402	3193	2350
	5					1866	814	2814	1762
	6					1487	226	2436	1173

Piston-Type Spring-Return

Use the following information to answer questions 1 and 2: a 3" sleeved plug valve with fail closed action, a 60 psig air supply, and a piston-type actuator.

_____ **1.** What is the required break torque, including the recommended safety factor?

_____ **2.** What is the proper model and size for the actuator?

Use the following information to answer questions 3 and 4: a 4" lined plug valve with fail open action, an 80 psig air supply, and a piston-type actuator.

_____ **3.** What is the required break torque, including the recommended safety factor?

_____ **4.** What is the proper model and size for the actuator?

Use the following information to answer questions 5 and 6: a 6" lined ball valve with fail closed action, a 100 psig air supply, and a piston-type actuator.

_____ **5.** What is the required break torque, including the recommended safety factor?

_____ **6.** What is the proper model and size for the actuator?

Use the following information to answer questions 7 and 8: a 10" sleeved plug valve with double-acting movement, a 100 psig air supply, and a piston-type actuator.

_____ **7.** What is the required break torque, including the recommended safety factor?

_____ **8.** What is the proper model and size for the actuator?

Actuators and Positioners
ACTIVITIES

Name _____ Date _____

Activity 41-4—Solenoid Selection

Given the valve descriptions in the questions and the solenoid catalog information shown below, select an appropriate solenoid for each valve. It should have the largest capacity with ⅜″ NPT connections, if available, while still being able to handle 100 psig air. It is generally recommended to select the solenoid with the largest orifice, connections, and tubing, so that the actuator air chamber can be vented as quickly as possible. If the valve opens or closes too fast, there are devices that can be added to the solenoid to slow the movement of air.

Three-Way Solenoid Specifications

Pipe Size (in.)	Orifice Size (in.)	C_V Flow Factor	Operating Pressure Differential (psi)						Max. Fluid Temp. °F		Standard Solenoid Enclosures				Watt Rating/ Class of Coil Insulation	
			Max. AC			Max. DC					Brass Body		S.S. Body			
			Air-Inert Gas	Water	Lt. Oil @ 300 SSU	Air-Inert Gas	Water	Lt. Oil @ 300 SSU	AC	DC	Catalog Number	Constr. Ref. No.	Catalog Number	Constr. Ref. No.	AC	DC
UNIVERSAL OPERATION (Pressure at any port)																
⅛	³⁄₆₄	0.06	175	175	175	125	125	125	140	120	8320G130	1	8320G140	1	9.1/F	10.6/F
⅛	¹⁄₁₆	0.09	100	100	100	65	65	65	180	120	8320G1	1	8320G41	1	9.1/F	10.6/F
⅛	³⁄₃₂	0.12	50	50	50	50	50	50	180	120	8320G83	1	8320G87	1	9.1/F	10.6/F
⅛	⅛	0.21	30	30	30	20	20	20	180	120	8320G3	1	8320G43	1	9.1/F	10.6/F
¼	¹⁄₁₆	0.09	125	130	130	75	75	75	200	150	8320G172	4			10.1/F	11.6/F
¼	³⁄₃₂	0.12	100	100	100	60	60	60	200	150	8320G174	4	8320G200	5	10.1/F	11.6/F
¼	⅛	0.25	50	50	50	25	25	25	200	150	8320G176	4	8320G201	5	10.1/F	11.6/F
¼	¹¹⁄₆₄	0.35	20	20	20	12	12	12	200	150	8320G178	4			10.1/F	11.6/F

1. A piston-type, 3″ sleeved plug valve, with fail closed action, and 60 psig air supply.

_____

2. A piston-type, 6″ lined plug valve, with fail closed action, and 80 psig air supply.

_____

Name _____ **Date** _____

Activity 41-5—Pneumatic Systems

1. The flapper/nozzle relationship is the most fundamental part of every pneumatic transmission signal source. Draw this relationship showing the air supply, restrictor, nozzle, flapper, and output air signal.

2. What is the important relationship between the nozzle and restrictor?

   _____

   _____

3. Add a cutaway view of a pneumatic relay and a feedback bellows to the drawing.

4. Describe how the pneumatic relay works.

   _____

   _____

5. The flapper/nozzle combined with the pneumatic relay comprises a high-gain amplifier. Describe the necessity of a feedback signal from the output and what it does.

   _____

   _____

Name _____ **Date** _____

Activity 41-6—Air Drying

The vapor pressure of a liquid is the pressure developed by the molecules of the liquid as they evaporate and leave the surface of the liquid to form vapor. The molecules continue leaving the surface of the liquid until there is a sufficient number in vapor form that there is an equilibrium between the number of molecules leaving the surface and the number of molecules hitting the surface and becoming liquid again. The number of liquid molecules in the vapor space depends on the temperature of the liquid and vapor. Typically the liquid and the vapor are at the same temperature.

Vapor pressure increases with increases in temperature. The vapor pressure of water at 212°F is 14.7 psia. The vapor pressure of water at 100°F is 0.949 psia. These values can be obtained from the saturated steam tables. When a liquid is in contact with gas, the vapor molecules from the liquid share the vapor space with the gas. The pressure due to the liquid's vapor molecules is called "partial pressure," where the remainder of the total pressure is the gas.

For example, hot water and air in a closed container could have the following characteristics:
Water temperature = 140°F
Water vapor pressure = 2.889 psia
Total pressure = 20.0 psia
Air partial pressure = 20.0 psia – 2.889 psia = 17.11 psia

The partial pressures in a gas mixture are equal to the partial volume of those components. The partial volumes are calculated as follows:

Partial volume water vapor = water partial pressure ÷ total pressure
2.889 psia (H_2O) ÷ 20.0 psia = 0.144
0.144 × 100 = **14.4% water vapor**

Partial volume air = air partial pressure ÷ total pressure
17.111 psia (air) ÷ 20.0 psia = 0.856
0.856 × 100 = **85.6% air**

The specific gravity of a gas/vapor mixture can be determined by multiplying each gas volume fraction (percentage ÷ 100) by its molecular weight with the results of all the individual gas/vapor components added together and then divided by 29, the molecular weight of air.

Example:

	Volume Fraction		Mol Wt	
Water vapor (H_2O)	0.144	×	18	= 2.59
Air	0.856	×	29	= 24.82
	Gas/vapor molecular weight			= 27.41

Specific gravity of gas/vapor mixture = 27.41 ÷ 29.0 = **0.945**

Instrument air is often dried to a dewpoint of –40°F. A heatless air dryer consists of two desiccant beds that are piped in parallel. Moist compressed air passes through one desiccant bed and the desiccant removes moisture from the air. Dry purge air is sent through the second bed to dry the desiccant in that bed. When the desiccant in the first bed is saturated, selector valves are switched to start sending the moist air through the second bed and purge air is sent through the first bed.

A flow of 50 scfm of compressed air at 100 psig and 100°F that is fully saturated with water vapor is sent through the air dryer. The exiting air has a dewpoint of –40°F.

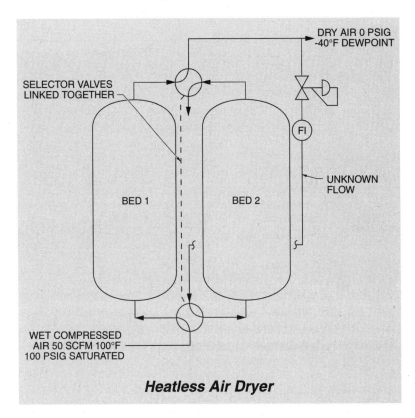

Heatless Air Dryer

_____ **1.** What is the flow rate of purge air through Bed 2?

_____ **2.** What percentage of the incoming air is available for use in the plant as instrument air with a –40°F dewpoint?

SECTION
10 FINAL ELEMENTS

chapter
42

Variable-Speed Drives and
Electric Power Controllers

REVIEW
QUESTIONS

Name _____   **Date** _____

_____ **1.** A ___ is a device that varies the speed of an electric motor.
 A. variable-speed drive
 B. triac
 C. regulator
 D. positioner

_____ **2.** An AC variable-speed drive changes the ___ of the voltage applied across the motor contacts to change the motor speed.
 A. phase
 B. frequency
 C. resistance
 D. power

_____ **3.** A DC variable-speed drive changes the ___ of the voltage applied across the motor contacts to change the motor speed.

_____ **4.** For a blower or fan, doubling the speed results in a ___ of the air flow.
 A. halving
 B. doubling
 C. tripling
 D. quadrupling

_____ **5.** For a pumping system, ___ is the change in elevation of the discharge piping system and remains constant for all flow rates.
 A. elevation change
 B. height increase
 C. fixed pressure
 D. static head

_____ **6.** For a pumping system, ___ is the pressure loss due to flow through the piping and varies with the flow squared.
 A. frictional head
 B. flow pressure
 C. dynamic flow
 D. Bernoulli exchange

_____ **7.** A ___ pump is used to provide a constant flow rate at any discharge pressure for a given operating speed.
 A. positive-displacement
 B. hydraulic-curve
 C. static-head
 D. dynamic-head

_____ **8.** A ___ pump is used to provide a constant discharge pressure at any flow rate for a given operating speed.
　　A. constant-volume
　　B. gear
　　C. centrifugal
　　D. dynamic-head

_____ **9.** A ___ is a plot of the pump discharge pressure against flow for various pump rotational speeds.
　　A. static-head graph
　　B. flow graph
　　C. pressure plot
　　D. pump curve

_____ **10.** A ___ is a flow curve that takes into account the static and frictional heads of the process.
　　A. static-head graph
　　B. frictional-head graph
　　C. hydraulic curve
　　D. friction-loss curve

_____ **11.** A ___ is a semiconductor switching device that uses a low-current DC input to switch an AC circuit.
　　A. mechanical switch
　　B. solid-state relay
　　C. direct controller
　　D. break

T　　F　　**12.** A zero switching relay is a solid-state relay where the load becomes energized as soon as the control input voltage is applied.

_____ **13.** A(n) ___ relay allows for proportional control and a ramp-up function of the load.
　　A. analog switching
　　B. zero switching
　　C. single-pole
　　D. double-throw

_____ **14.** A ___ is a solid-state switching device that is used to switch very-low-current DC loads.
　　A. break
　　B. control relay
　　C. zero switch
　　D. transistor

_____ **15.** A(n) ___ is a solid-state power controller that provides a proportional current to a heating element in response to an analog control signal.
　　A. zero switching relay (ZSR)
　　B. transistor switched controller (TSC)
　　C. silicon controlled rectifier (SCR)
　　D. analog switching relay (ASR)

Name _____   Date _____

Activity 42-1—Drive Selection

In the following questions you will be asked to determine whether the application is suitable for a variable-speed drive on the pump instead of a control valve. Some applications will be obvious, some are not obvious, and some may require a hydraulic analysis. If there is insufficient information, state that fact. In all cases, state the reason for your decision.

A control valve is installed in a long 3″ pipeline where, at a design flow of 100 gpm, the pipeline $\Delta p = 43$ psig and the control valve $\Delta p = 7$ psig. The calculated control valve size is $C_v = 39$ at design conditions. At minimum flow of 10 gpm, the calculated C_v is 1.5.

1. Is this application suitable for a variable-speed drive?

_____

_____

_____

Hot condensate at 220°F from a collection tank is pumped up 20 ft through a control valve into a deaerator. The flow out of the pump is 15 gpm at 25 psig and 220°F. There is always the possibility of flashing when handling hot condensate. Flashing limits the Δp to the difference between the fluid vapor pressure and the inlet pressure.

2. Is this application suitable for a variable-speed drive?

_____

_____

_____

The primary steam feed to a multistage evaporator uses 250 psig saturated steam. It is fed to the primary steam heat exchanger that transfers heat to the most concentrated solution in the evaporator. Under normal operating conditions, the pressure in the steam chest is 180 psig with a control valve $\Delta p = 70$ psig.

3. Is this application suitable for a variable-speed drive?

_____

_____

_____

A flow control loop feeds 31% HCl acid to a steel pickling bath. The flow range is 0 gpm to 15 gpm and the design flow is 10.0 gpm. The design flow $C_v = 3.45$. The piping size is 1½" TFE lined pipe.

4. Is this application suitable for a variable-speed drive?

_____

_____

_____

Cooling water is fed to a heat exchanger through a 6" pipe. The design $C_v = 104$ and the control valve is only required to have a 4:1 turndown.

5. Is this application suitable for a variable-speed drive?

_____

_____

_____

A waste treatment plant requires the use of 15% caustic hydroxide solution for the neutralization of acid wastes. The design case requires a $C_v = 0.041$.

6. Is this application suitable for a variable-speed drive?

_____

_____

_____

INTEGRATED
ACTIVITY

Name _____ **Date** _____

Piping Resistance

A control valve is installed in a long 3″ pipeline where, at a design flow of 100 gpm, the pipeline $\Delta p = 43$ psig and the control valve $\Delta p = 7$ psig. The calculated control valve size is $C_v = 39$ at design conditions. At minimum flow of 10 gpm, the calculated C_v is 1.5. This control valve is used in a long pipeline where the large majority of the pressure drop is due to piping frictional losses. The selected valve is a 2″ equal-percentage valve. The actual flow through the valve will change from the equal-percentage characteristic because of the effect of the piping resistance.

% Travel	Catalog C_V
100	56.2
90	47.8
80	36.7
70	25.0
60	16.3
50	10.6
40	3.91
30	4.72
20	3.15
10	1.74
0	0

Use the basic equation of flow, flow = $C_p\sqrt{\Delta p}$, to complete the following activity. Keep in mind that there are two equations for flow, one for the pipeline and the other for the valve, and that the flow and appropriate changes in pressure must satisfy both equations.

1. Use the graph paper on the back of this page and plot the catalog C_v as a function of percent travel for the valve.

2. On the same graph as question 1, plot the changed flow characteristic due to the piping resistance.

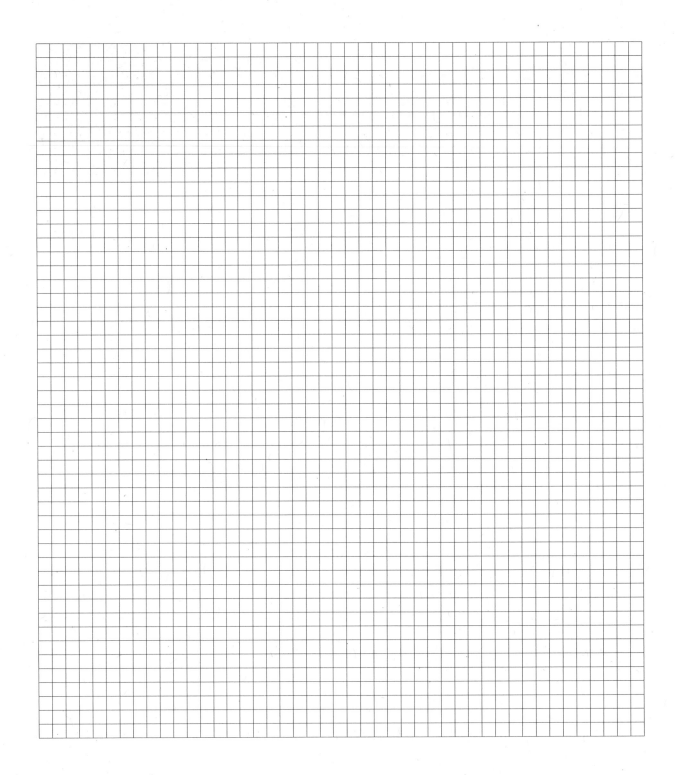

SECTION
11 SAFETY SYSTEMS

chapter
43
Safety Devices and Equipment

REVIEW
QUESTIONS

Name _____   **Date** _____

_____ 1. ___ is any clothing or device worn by a worker to prevent injury.

_____ 2. Safety valves, relief valves, and safety relief valves must conform to codes established by ___.
   A. ISA
   B. ASME International
   C. IEEE
   D. OSHA

_____ 3. A(n) ___ is a recessed area in a safety valve disc that increases the surface area and the total force applied.

_____ 4. A(n) ___ is a valve that opens in proportion to the pressure above the setpoint.

_____ 5. A safety relief valve has a(n) ___, sometimes called a blowdown ring, that is used to adjust the valve opening speed.
   A. adjustment chamber
   B. control ring
   C. setpoint ring
   D. huddling chamber

_____ 6. A(n) ___ is a safety device used to prevent damage in pipelines and pressure vessels due to excessive pressure.
   A. open vent
   B. backpressure regulator
   C. safety disc
   D. rupture disc

_____ 7. A(n) ___ is a specially designed solid-state electronics package that provides the sequencing and safety logic for controlling industrial burners.

_____ 8. A(n) ___ is a device that can provide a usable signal to a burner control system by detecting the presence of a flame.
   A. flame detector
   B. ionization sensor
   C. oxygen sensor
   D. fuel flow switch

_____ 9. A(n) ___ is a solid-state or relay-based system for monitoring and alarming plant process operations.

_____ 10. A(n) ___ is a feature that is available in annunciators and digital alarm systems that can identify the first of multiple closely tripped alarms.

_____ 11. A(n) ___ is a hazardous atmosphere detector used to measure low concentrations of combustible gases and vapors in the atmosphere.
     A. upper explosive limit
     B. lower explosive limit
     C. flame sensor
     D. combustible gas detector

_____ 12. The safety valve capacity of a boiler safety valve is the amount of ___, in lb/hr, that the valve is capable of venting at the rated pressure.
     A. steam
     B. air
     C. fuel
     D. gas

_____ 13. A(n) ___ valve is a device that shuts off the flow if the flow exceeds a specific flow rate.

_____ 14. ___ safety relief valves are approved for use on tank cars and tank trucks where standard safety valves are not approved for these services.
     A. Hazardous material
     B. Chemical transportation
     C. Dangerous fluid
     D. EPA-approved

_____ 15. A runaway chemical reaction can occur due to ___ or due to a reaction rate exceeding the cooling capacity.
     A. excess cooling
     B. thermal expansion
     C. loss of cooling
     D. blocked discharge

_____ 16. A ___ valve is a special spring-actuated valve used to stop the fuel flow to a burner system.
     A. fuel flow control
     B. safety relief
     C. rupture pin safety
     D. fuel safety shutoff

_____ 17. A(n) ___ detector is used to detect small traces of hazardous gases when they are at concentrations too low to ignite or below harmful levels.

_____ 18. A(n) ___ detector is an instrument used to measure poisonous gases or vapors that are not combustible but are harmful to people.

Name _____   Date _____

Activity 43-1—Worst-Case Scenarios

Safety relief devices are sized to handle the worst possible condition. The worst possible condition is the one that requires the greatest amount of vapor, gas, or liquid to be released. There are four conditions that need to be examined to determine which is the worst case. They are fire, run-away chemical reaction, blocked discharge, and thermal expansion. It is not necessary to calculate all of the conditions for every application. An examination of the system can eliminate some of the conditions.

In the following examples, state which condition or conditions need to be examined.

1. Heat exchanger pressure safety valve (PSV)

   _____

   _____

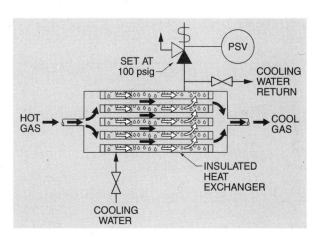

2. Regulator discharge pressure safety valve (PSV)

   _____

   _____

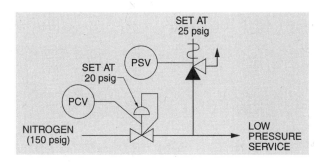

3. LPG storage pressure safety valve (PSV)

_____

_____

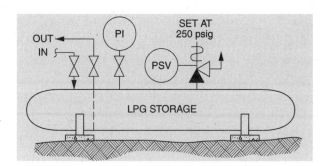

4. Reactor vessel pressure safety valve (PSV 4)

_____

_____

5. Reactor jacket pressure safety valve (PSV 5)

_____

_____

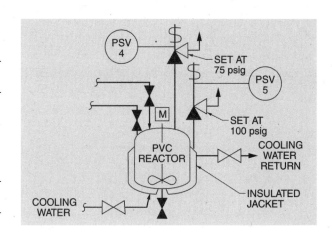

6. Hexane tank pressure safety valve (PSV 6)

_____

_____

7. Hexane tank vacuum relief safety valve (PSV 7)

_____

_____

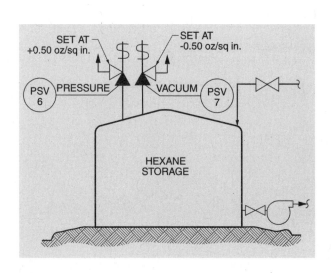

Name _____   Date _____

Activity 43-2—Relief Requirements—Air

A pressure reducing valve is set at 10 psig and sized to handle 20 scfm of air with a 10% offset. The air supply is 100 psig. Use the equation $F_{scfh} = 39.5 \times C_v \times \sqrt{\Delta P \times P_{INa}}$ for calculating the maximum flow through the valve. The C_v for the valve at maximum flow is 4.4. ΔP is the pressure drop across the valve and P_{INa} is the inlet pressure to the valve in absolute terms. There is a pressure safety valve (PSV) downstream of the regulator set at a relief pressure of 15 psig that is to be sized for a 10% accumulation.

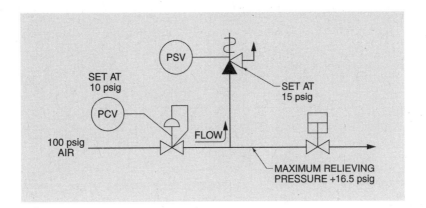

_____   **1.** What is the flow of air through the safety valve if the pressure control valve fails 100% open?

Name _____   Date _____

Activity 43-3—Relief Requirements—Steam

A natural gas-fired boiler is designed to generate 10,000 lb/hr of saturated steam at 135 psig. The feedwater is delivered at 60°F, and the 100% open natural gas control valve will pass a maximum flow of 393 scfm. Assume that the natural gas heating value is 1050 Btu/cu ft and that 100% of this heat is used. The boiler has a safety valve set at 160 psig and should be sized to reach its required capacity at 3% overpressure.

Use the steam table in the Appendix to answer the following question.

_____   **1.** What is the required steam flow rate through the safety valve?

SECTION

11 *SAFETY SYSTEMS*

Safety Devices and Equipment

chapter

43

ACTIVITIES

Name _____ **Date** _____

Activity 43-4—Relief Requirements—LPG

Liquefied petroleum gas (LPG) is compressed and stored in a pressurized vessel. The storage container is a horizontal cylindrical vessel with hemispherical ends, 10′ in diameter, and 60′ in overall length.

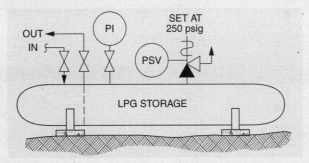

The only case that needs to be checked is fire. For uninsulated containers for liquefied gases, the minimum required flow capacity of the safety relief device shall be calculated using the following formula:

$$Q_a = G_n \times A^{0.82}$$

where

Q_a = flow capacity in scfm of free air

G_n = gas factor for uninsulated containers

A = total outside surface area of the container in sq ft

For liquefied petroleum gas, the value of G_n is 53.6. Values for other gases can be obtained from manufacturers of safety relief valves.

When the surface area is not stamped on the nameplate, the area can be calculated by using one of the following formulas:

1. Cylindrical container with hemispherical heads:
    A = (overall length) × (outside diameter) × π

2. Spherical container:
    A = (outside diameter)^2 × π
    where
    A = area
    π = 3.14

257

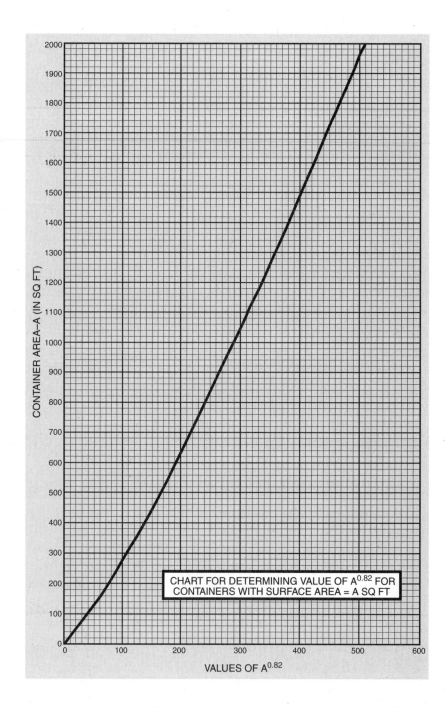

CHART FOR DETERMINING VALUE OF $A^{0.82}$ FOR CONTAINERS WITH SURFACE AREA = A SQ FT

Use the chart on this page or use a scientific calculator to determine the value of $A^{0.82}$ for the formula.

_____ **1.** What is the vapor release rate for which the valve must be sized?

_____ **2.** What is the relieving pressure for which the valve must be sized with a 10% accumulation?

Name _____   Date _____

Activity 43-5—Relief Requirements—Atmospheric Storage Tank

An atmospheric cylindrical vertical storage tank containing hexane is 30 ft in diameter, with a 15 ft vertical wall. Hexane has a heat of vaporization of 144 Btu/lb and a molecular weight of 86.17.

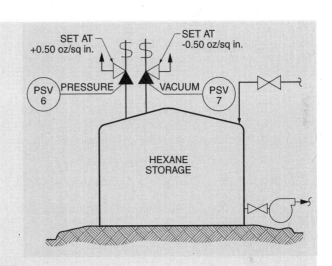

Use the following formula to determine the venting requirement:

$$CFH = \frac{70.5 \times Q}{L \times \sqrt{M}}$$

where
CFH = venting requirement in scfh of free air
Q = total heat input in Btu/hr
L = latent heat of vaporization
M = molecular weight

Use the chart on the back of this page or use one of the following formulas to determine the amount of heat transferred into the tank. Note that the equations provide a more accurate answer because the graph is difficult to read accurately.

$Q = 20,000 \times A$ for $A < 200$ sq ft
$Q = 199,300 \times A^{0.566}$ for A between 200 sq ft and 1000 sq ft
$Q = 963,400 \times A^{0.338}$ for A between 1000 sq ft and 2800 sq ft

where
Q = heat transfer in Btu/hr
A = area in sq ft

_____ **1.** What is the equivalent atmospheric air flow required?

_____ **2.** If the discharge pump has a maximum pumping rate of 36 gpm, what vacuum relief capacity in cu ft/hr is required?

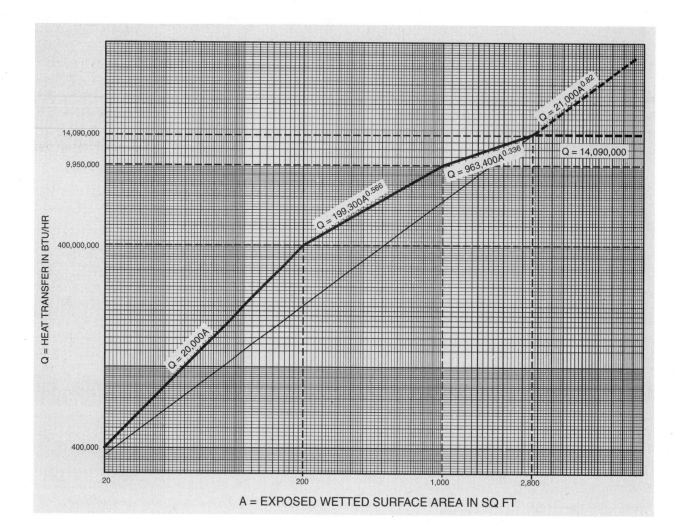

Safety Devices and Equipment

ACTIVITIES

Name _____ Date _____

Activity 43-6—Safety Valve Selection—Air

A safety valve is used to protect equipment down-stream of a pressure-reducing valve. The safety valve must be selected to be able to relieve 283 scfm of air at 15 psig.

Safety Relief Valve Sizes

Nozzle Area*	Safety Relief Valve Sizes (Inlet†, Nozzle Code, Outlet†)		
0.110	1	D	2
0.196	1	E	2
0.307	1½	F	2
0.503	1½	G	2½
0.785	1½	H	3
1.287	2	J	3
1.838	3	K	4
2.853	3	L	4
3.60	4	M	6
4.34	4	N	6
6.38	4	P	6
11.05	6	Q	8
16.0	6	R	8
26.0	8	T	10

* in sq in.
† in in.

Air Capacities*

Set Pressure (psig)	Orifice Letter and Effective Area, Sq In.													
	D 0.110	E 0.196	F 0.307	G 0.503	H 0.785	J 1.287	K 1.838	L 2.853	M 3.60	N 4.34	P 6.38	Q 11.05	R 16.0	T 26.0
1 psi inc.	2	4	6	10	15.5	25	36	56	70	85	125	217	314	511
5 psi inc.	11	19	30	49	77	126	180	280	354	426	627	1086	1573	2557
15	64	115	179	294	459	752	1075	1668	2105	2537	3730	6461	9355	15202
20	74	132	207	339	529	858	1239	1923	2427	2925	4301	7449	10785	17526
30	94	167	262	429	670	1098	1568	2433	3070	3701	5441	9424	13646	22175
40	115	206	322	528	824	1351	1929	2994	3778	4555	6696	11598	16793	27288
50	137	244	383	627	978	1604	2291	3556	4486	5409	7951	13771	19940	32402

* in scfm at 60°F and 10% overpressurev

Use the tables to answer the following questions.

_____ 1. What is the appropriate orifice letter designation? If it requires more than one relief valve, select equal-size valves.

_____ 2. What is the effective area of the orifice chosen in Question 1?

_____ 3. What are the appropriate inlet and outlet safety valve nozzle sizes?

_____ 4. Suggest changes in the system that can be made to reduce the relieving capacity.

Safety Devices and Equipment

Name _____   Date _____

Activity 43-7—Safety Valve Selection—Steam

A safety valve is used to protect a boiler from overpressure. The safety valve must be sized to relieve 21,270 lb/hr of steam at 175 psig.

Safety Relief Valve Sizes

Nozzle Area*	Safety Relief Valve Sizes (Inlet†, Nozzle Code, Outlet†)		
0.110	1	D	2
0.196	1	E	2
0.307	1½	F	2
0.503	1½	G	2½
0.785	1½	H	3
1.287	2	J	3
1.838	3	K	4
2.853	3	L	4
3.60	4	M	6
4.34	4	N	6
6.38	4	P	6
11.05	6	Q	8
16.0	6	R	8
26.0	8	T	10

* in sq in.
† in in.

Saturated Steam Capacities†

Set Pressure (psig)	Orifice Letter and Effective Area, Sq In.					
	G 0.503	H 0.785	J 1.287	K 1.838	L 3.60	M 3.60
110	3413	5326	8732	12471	19358	24426
120	3690	5758	9440	13482	20927	26406
130	3966	6190	10148	14493	22496	28386
140	4243	6621	10856	15504	24065	30366
150	4519	7053	11564	16514	25634	32346
160	4796	7485	12272	17525	27203	34326
170	5073	7917	12979	18536	28773	36306
180	5349	8348	13687	19547	30342	38286
190	5626	8780	14395	20558	31911	40266
200	5903	9212	15103	21569	33480	42246

† in lb/hr at 10% overpressure

Use the tables to answer the following questions.

_____ **1.** What is the appropriate orifice letter designation? If it requires more than one relief valve, select equal-size valves.

_____ **2.** What is the effective area of the orifice chosen in Question 1?

_____ **3.** What are the appropriate inlet and outlet safety valve nozzle sizes?

Safety Devices and Equipment

ACTIVITIES

Name _____ **Date** _____

Activity 43-8—Safety Valve Selection—LPG

A safety valve is used to protect an LPG storage tank from overpressure due to fire. The safety valve must be sized to relieve 25,996 scfm air equivalent at 250 psig. The valve was sized for a 10% accumulation. The LPG gas is propane. The temperature of the propane with a vapor pressure of 275 psig is about 132°F.

Safety Relief Valve Sizes

Nozzle Area*	Safety Relief Valve Sizes (Inlet†, Nozzle Code, Outlet†)		
0.110	1	D	2
0.196	1	E	2
0.307	1½	F	2
0.503	1½	G	2½
0.785	1½	H	3
1.287	2	J	3
1.838	3	K	4
2.853	3	L	4
3.60	4	M	6
4.34	4	N	6
6.38	4	P	6
11.05	6	Q	8
16.0	6	R	8
26.0	8	T	10

* in sq in.
† in in.

Air Capacities†

Set Pressure (psig)	Orifice Letter and Effective Area, Sq In.													
	D 0.110	E 0.196	F 0.307	G 0.503	H 0.785	J 1.287	K 1.838	L 2.853	M 3.60	N 4.34	P 6.38	Q 11.05	R 16.0	T 26.0
*1 psi inc.	2	4	6	10	15.5	25	36	56	70	85	125	217	314	511
*5 psi inc.	11	19	30	49	77	126	180	280	354	426	627	1086	1573	2557
220	505	900	1409	2309	3603	5907	8436	13095	16523	19920	29283	50717	73437	119335
240	548	977	1530	2507	3912	6413	9159	14217	17939	21627	31793	55064	79730	129562
260	591	1054	1651	2704	4221	6920	9882	15339	19355	23334	34308	59411	86024	139789
280	635	1131	1771	2902	4529	7426	10605	16461	20772	25041	36812	63757	92318	150017
300	678	1208	1892	3100	4838	7932	11328	17584	22188	26748	39321	68104	98612	160244

Use the tables to answer the following questions.

_____ 1. What is the appropriate orifice letter designation? If it requires more than one relief valve, select equal-size valves.

_____ 2. What is the effective area of the orifice chosen in Question 1?

_____ 3. What are the appropriate inlet and outlet safety valve nozzle sizes?

_____ 4. What could be done to reduce the safety valve size requirement?

Safety Devices and Equipment

Name _____   Date _____

Activity 43-9—Safety Valve Selection—Atmospheric Storage Tank

An atmospheric storage tank has a breather valve to protect it from pressure. The tank has a relieving pressure of 10″ WC with a set pressure of ±0.5 oz/sq in. The safety valve is required to relieve 550,000 scfh of air. The breather valve opens to relieve the vacuum that is created as liquid in the tank is pumped out. The vacuum relief capacity is 289 scfh air. Atmospheric breather relief valves are constructed with one flow channel to allow a pressurized fluid to escape and another flow channel to allow air to flow into the tank to relieve a vacuum.

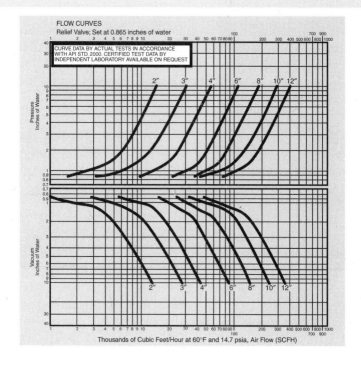

FLOW CURVES
Relief Valve; Set at 0.865 inches of water

Thousands of Cubic Feet/Hour at 60°F and 14.7 psia, Air Flow (SCFH)

Use the chart to answer the following questions.

_____   **1.** What is the required valve size for pressure relief? If it requires more than one relief valve, select equal-size valves.

_____   **2.** What is the relieving vacuum at maximum relief rate?

Name _____   Date _____

Activity 43-10—Alarms

The piping and instrumentation diagram (P&ID) on the back of this page shows a boiler system fired by two fuels, natural gas and hydrogen. Each fuel can be operated independently or with the other fuel. The loss of one fuel train will not affect the other. The pilot system shuts down after the main flame is confirmed. There are four interlocking systems with associated alarm points plus some extra alarms. The following is a list of alarms associated with the burner system:

BAL-12	Burner system, OFF
LAH-06	Boiler water level, High
LAL-05	Boiler water level, Low
LALL-07	Boiler water level, Low Low
LALL-08	Boiler water level, Low Low
PAH-10	Boiler pressure, High
PAHH-11	Boiler pressure, High High
PAH-21	Natural gas burner pressure, High
PAH-22	Hydrogen gas burner pressure, High
PAL-01	Combustion air blower, OFF
PAL-15	Natural gas supply pressure, Low
PAL-19	Hydrogen gas supply pressure, Low

A first-out group is a group of alarms that are linked together so that only the first alarm tripped activates the alarm. The purpose of a first-out group is to help operators understand the true initiating cause of a set of alarms.

Use the list of alarms and the P&ID to perform the following task.

1. Arrange the alarms into a number of groups that should be grouped as first-out groups.

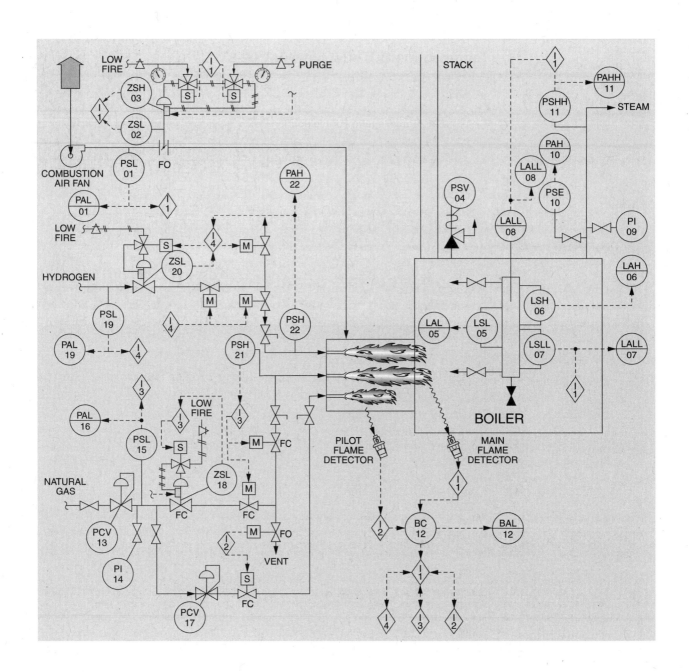

Name _____   **Date** _____

Activity 43-11—Rupture Disc Selection—Steam

A heat exchanger that cools process gases uses cooling water on the shell side of the shell and tube exchanger. The shell of the exchanger is 10 ft long and 3 ft in diameter, and is not insulated. The design pressure of shell and tubes is 100 psig. Use one of the following formulas to determine the amount of heat transferred into the tank.

$Q = 20,000 \times A$ for $A < 200$ sq ft
$Q = 199,300 \times A^{0.566}$ for A between 200 sq ft and 1000 sq ft
$Q = 963,400 \times A^{0.338}$ for A between 1000 sq ft and 2800 sq ft
$Q = 21,000 \times A^{0.82}$ for A between 2800 sq ft and 10,000 sq ft

where
Q = heat transfer in Btu/hr
A = area in sq ft

	Rupture Disc Selection Table										
	Rupture Pressure (psig)	**Dry and Saturated Steam: Pounds per Hour**									
		½	1	1½	2	3	4	6	8	10	12	16
Steam	**50**	390	1,520	3,450	6,100	13,800	24,600	55,000	98,000	152,000	220,000	370,000
	100	690	2,700	6,100	10,800	24,000	43,400	97,000	174,000	270,000	390,000	650,000
	200	1,290	5,040	11,400	20,200	45,600	81,000	182,000	325,000	504,000	730,000	1,200,000
	400	2,490	9,700	22,000	39,100	88,000	157,000	352,000	620,000	970,000	—	—
	600	3,690	14,400	32,700	58,000	130,000	233,000	523,000	930,000	1,400,000	—	—
	800	4,900	19,100	43,300	77,000	173,000	308,000	690,000	1,200,000	—	—	—
	1,000	6,100	23,700	53,800	95,000	215,000	383,000	860,000	1,500,000	—	—	—
	1,500	9,100	35,500	80,000	142,000	320,000	570,000	1,300,000	—	—	—	—
	2,000	12,100	47,200	170,000	190,000	430,000	—	—	—	—	—	—
	2,500	15,100	59,000	134,000	237,000	—	—	—	—	—	—	—
	3,000	18,100	—	—	—	—	—	—	—	—	—	—

Use the formulas, the steam table in the Appendix, and the rupture disc selection table to answer the following questions.

_____ **1.** What is the area of the heat exchanger potentially exposed to a fire?

_____ **2.** What is the potential amount of heat transferred into the heat exchanger?

_____ **3.** What is the required steam relieving capacity for a fire case?

_____ **4.** What is the rupture disc size and capacity?

SECTION
11 SAFETY SYSTEMS

Electrical Safety Standards

chapter
44

REVIEW
QUESTIONS

Name _____ **Date** _____

T   F   **1.** The primary function of the National Electrical Code® is to safeguard people and property against hazards due to boiler explosions.

_____   **2.** Electrical classifications for hazardous locations are identified by class, division, and ___.

_____   **3.** A Class ___ location is a hazardous location in which sufficient quantities of combustible dust are present in the air to cause an explosion or to ignite the hazardous material.
   A. I
   B. II
   C. III
   D. IV

_____   **4.** A ___ is the classification assigned to each class based upon the likelihood of the presence of the hazardous substance in the atmosphere.
   A. group
   B. class
   C. location
   D. division

T   F   **5.** A Division 2 location is a hazardous location in which the hazardous substance is normally present in the air in sufficient quantities to cause an explosion or to ignite the hazardous materials.

_____   **6.** An enclosure ___ is a designation that specifies the maximum surface skin temperature obtained by a piece of equipment during testing by approval agencies.

T   F   **7.** Explosionproof protection is a type of protection that excludes ignitable amounts of dust or amounts that can affect performance or rating.

_____   **8.** A(n) ___ is the case or housing of equipment or other apparatus that provides protection for controllers, motor drives, or other devices.
   A. junction box
   B. electrical safety box
   C. enclosure
   D. wiring protection device

_____ **9.** A common enclosure for Class I hazardous locations is a Type ___ enclosure.
     A. 4
     B. 4X
     C. 6
     D. 7

_____ **10.** A(n) ___ enclosure is a nonhazardous enclosure that is pressurized and purged with air to allow it to be used in hazardous areas.

_____ **11.** The wiring requirements in hazardous areas are defined by ___.
     A. OSHA
     B. NEC®
     C. NEMA®
     D. IEEE

_____ **12.** ___ protection is where the electrical equipment, under normal or abnormal conditions, is incapable of releasing sufficient electrical or thermal energy to cause ignition.
     A. Intrinsically safe
     B. Nonincendive
     C. Ignition-proof
     D. Explosionproof

_____ **13.** A(n) ___ is a specially designed electronic circuit containing resistors and diodes which is used to prevent any electrical ignition energy from being carried into a hazardous area.

Electrical Safety Standards

ACTIVITIES

44

Name _____   Date _____

Activity 44-1—Hazardous Atmospheres

The following table shows the low flammable limit (LFL) and high flammable limit (HFL) for several common flammable gases.

Flammable Material	LFL	HFL
Hydrogen	4.0%	74.2%
Acetylene	2.5%	80.0%
Methane	5.0%	15.0%
Propane	2.1%	10.1%
Natural gas	4.9%	15.0%
Gasoline	1.4% to 1.5%	7.4% to 7.6%

Flammable gases are listed below along with an air:gas ratio. Use that information along with the above table to answer the questions.

Natural gas – 20:1

_____ **1.** What is the percent gas in the mixture?

_____ **2.** Is the mixture flammable?

Methane – 5:1

_____ **3.** What is the percent gas in the mixture?

_____ **4.** Is the mixture flammable?

Hydrogen gas – 1:2

_____ **5.** What is the percent gas in the mixture?

_____ **6.** Is the mixture flammable?

Propane gas – 10:1

_____ **7.** What is the percent gas in the mixture?

_____ **8.** Is the mixture flammable?

Gasoline – 20:1

_____ **9.** What is the percent gas in the mixture?

_____ **10.** Is the mixture flammable?

Acetylene gas – 1:10

_____ **11.** What is the percent gas in the mixture?

_____ **12.** Is the mixture flammable?

Propane gas – 50:1

_____ **13.** What is the percent gas in the mixture?

_____ **14.** Is the mixture flammable?

Natural gas – 10:1

_____ **15.** What is the percent gas in the mixture?

_____ **16.** Is the mixture flammable?

© American Technical Publishers, Inc.
All rights reserved

271

SECTION
11 SAFETY SYSTEMS

chapter
44

Electrical Safety Standards

ACTIVITIES

Name _____ Date _____

Activity 44-2—Purged Cabinets

An electrical cabinet 6′ high × 36″ wide × 18″ deep must be supplied with a rapid purge rate when the cabinet is initially closed after being open. The rapid purge rate must displace four volumes of the cabinet size in 2 min. During the high purge rate the air is carried out of the cabinet through a safety valve set at 0.3″ WC. The cabinet maintains a 0.1″ WC pressure with a 0.67 scfm air flow during normal purging rates.

Use the equation $F = C \times \sqrt{\Delta P}$ to calculate air flow.

_____ 1. What low flow rate is necessary to develop 0.2″ WC pressure after the initial high flow purge?

_____ 2. What rapid purge rate is needed?

3. Explain the difference in design between a purged cabinet in a Class I Division 1 environment and a purged cabinet in a Class I Division 2 environment.

_____

_____

_____

SECTION
11 SAFETY SYSTEMS

chapter
45

Safety Instrumented Systems

REVIEW
QUESTIONS

Name _____   **Date** _____

_____ **1.** A safety instrumented system (SIS) is a system consisting of sensors, logic solvers, and final control elements that bring a process to a safe state when ___.
  A. normal operating conditions are violated
  B. an emergency stop is actuated
  C. an operator shuts down a process
  D. all of the above

_____ **2.** The Occupational Safety and Health Administration has publicly identified the ___ standard as a "good engineering practice" when following OSHA regulations on the management of highly hazardous chemicals.
  A. OSHA CFR 1910
  B. EPA CFR 40
  C. ISA S84
  D. ISO 1994

_____ **3.** Risk is used to evaluate ___ and the consequences if a failure does happen.
  A. the severity of a failure
  B. how often a failure can happen
  C. the integrity levels
  D. system monitoring

_____ **4.** One common theme in all safety systems is to have ___, which increases the integrity level and reduces the residual risk.
  A. multiple levels of protection
  B. safety PLCs
  C. dual PLCs
  D. wireless systems

_____ **5.** A(n) ___ incident is one that could cause any serious injury or fatality on-site or property damage of $1 million off-site or $5 million on-site.
  A. minor
  B. serious
  C. severe
  D. unlikely

_____ **6.** A(n) ___ likelihood means that a failure, or series of failures, has a low probability of occurrence within the expected lifetime of the plant (1 in 100 to 1 in 10,000 per year).
  A. serious
  B. improbable
  C. occasional
  D. frequent

_____ 7. The IEC, ISA, and AIChE define integrity levels in terms of ___, probability of failure on demand (PFD), and risk reduction factor (RRF).
    A. safety availability
    B. severity
    C. ISA S84
    D. control

_____ 8. A(n) ___ is a highly reliable PLC that includes fail-safe designs, built-in self-diagnostics, and a fault-tolerant architecture.
    A. safety PLC
    B. nonconventional PLC
    C. polling controller
    D. logic solver

_____ 9. With ___ polling, redundant tripping signals from dual PLCs are needed to shut down a system.
    A. 1oo2
    B. 2oo2
    C. 1oo2D
    D. 2oo2D

T   F   **10.** System monitoring is the process of providing notification to an operator that there has been a failure in the process controlled by an SIS.

Name _____ **Date** _____

Hazardous Area Wiring

Two explosionproof instruments share a single branch conduit to a common junction box in a Class I Division 1 area.

1. Draw the required conduit details.

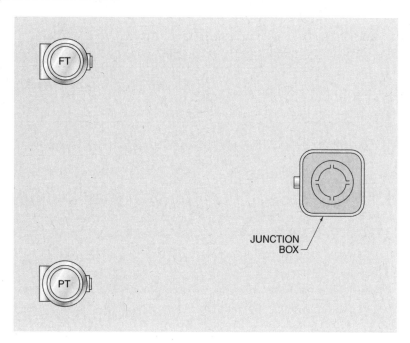

JUNCTION
BOX

SECTION
12 INSTRUMENTATION AND CONTROL APPLICATIONS

chapter
46

General Control Techniques

REVIEW
QUESTIONS

Name _____ Date _____

T   F   **1.** Two common types of operations are continuous and batch processes.

_____   **2.** A(n) ___ is a type of processing operation in which multiple feeds are received and the resulting product is fed all at once to the next operation.

_____   **3.** A(n) ___ control valve is an application where there are two or more control valves which operate using the same controller output signal range, with each valve using a portion of the controller output signal.
    A. split range
    B. equal-percentage
    C. shared output
    D. continuous

_____   **4.** Heating control valves are usually selected to have a fail closed action, and cooling control valves are usually selected to have a(n) ___ action.

T   F   **5.** Split range control valves can also be used to establish the order in which control valves are operated, and thus the priority between different operations.

_____   **6.** A ___ is a device that compares two or three input signals and passes the highest or lowest signal to the output of the device.
    A. divider
    B. multiplier
    C. splitter
    D. selector

_____   **7.** A ___ controller is used for those applications where control action is only desired above or below predetermined values.
    A. gap action
    B. proportional
    C. reverse-acting
    D. direct-acting

_____   **8.** A ___ controller is the arrangement of two controllers where a primary controller adjusts the setpoint of a secondary controller.
    A. ratio
    B. PID
    C. cascade
    D. gap action

_____ **9.** A(n) ___ selector can be used to select the highest of a group of measurements
to be used as the controller input.
   A. high
   B. low
   C. high/low
   D. ON/OFF

_____ **10.** ___ is the condition where a controller continues to change its output until the
output reaches its limit.

Name _____ Date _____

Activity 46-1—Split Range Valves—Fail-Safe Actions

A jacketed batch reactor is heated from a 550°F hot oil supply and is cooled with a water-cooled heat exchanger while the oil is being recycled. The reactor batch heat-up is programmed to take 2 hours. The batch is kept at 500°F for 6 hours and is then cooled down to 100°F in 8 hours. At this point the batch is done. The reactor is emptied and recharged for the next batch. The control signal is 4 mA to 20 mA and the control valves operate at 3 psig to 15 psig.

This application uses split range valves. During the heating phase of the batch cycle, the hot oil control valve, TV-1 01, is throttling, the recycle control valve, TV-3 01, is fully open, and the cooling control valve, TV-2 01, is fully closed. During the cooling phase of the batch cycle, the cooling control valve and the recycle control valve act in opposite directions during the heating phase. The safest failure action is with the cooling system completely ON and the heating system completely OFF.

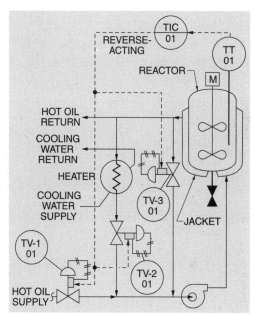

Use the P&ID to answer the following questions.

_____ 1. Should the hot oil valve, TV-1 01, be fail open or fail closed?

_____ 2. Should the cooling valve, TV-2 01, be fail open or fail closed?

_____ 3. Should the recycle valve, TV-3 01, be fail open or fail closed?

_____ 4. What is the desired portion of the control signal input to the valve positioner for the hot oil valve, TV-1 01?

_____ 5. What is the desired portion of the control signal input to the valve positioner for the cooling valve, TV-2 01?

_____ 6. What is the desired portion of the control signal input to the valve positioner for the recycle valve, TV-3 01?

_____ 7. What is the output of the valve positioner for the hot oil valve, TV-1 01?

_____ 8. What is the output of the valve positioner for the cooling valve, TV-2 01?

_____ 9. What is the output of the valve positioner for the recycle valve, TV-3 01?

Name _____   **Date** _____

Activity 46-2—Split Range Valves—Compressor Discharge Vent

The discharge pressure of a compressor is controlled by two split range valves, a vent valve, PV-1 02, and a user valve, PV-2 02. The vent valve is used to eliminate excess gas. The user valve is throttled to control the amount of gas being drawn by the user. The safest failure condition is to have all the gas sent to the vent. The control signal is 4 mA to 20 mA and the control valves operate at 3 psig to 15 psig.

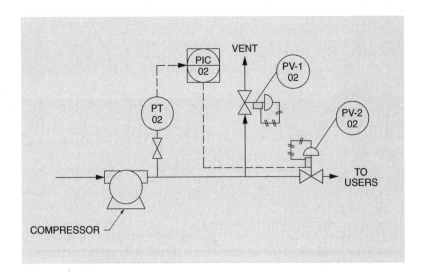

Use the P&ID to answer the following questions.

_____   **1.** Should the vent valve, PV-1 02, be fail open or fail closed?

_____   **2.** Should the user valve, PV-2 02, be fail open or fail closed?

_____   **3.** Should the controller, PIC 02, be direct acting or reverse acting?

_____   **4.** What is the desired portion of the control signal input to the valve positioner for the vent valve, PV-1 02?

_____   **5.** What is the desired portion of the control signal input to the valve positioner for the user valve, PV-2 02?

_____   **6.** What is the output of the valve positioner for the vent valve, PV-1 02?

_____   **7.** What is the output of the valve positioner for the user valve, PV-2 02?

Name _____ Date _____

Activity 46-3—Split Range Valves—Flow Splitting

The flow characteristics of some valves can be linearized from an equal-percentage characteristic with the use of a quick-opening cam in a valve positioner. The resulting characteristic curve can be determined by averaging the two characteristic curves over the percent travel.

A common application is the splitting of a flow stream in two directions depending on the value of some control variable. The two valves act in opposite directions. When one valve is closing, the other one is opening. A much more stable control can be obtained if the sum of the flows through the two valves is constant with any positions of the two valves.

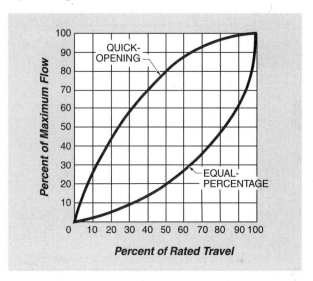

Use the equal-percentage and quick-opening curves to complete the following tasks.

1. Calculate the resulting linearized curve when using a quick-opening cam on an equal-percentage valve. Plot the result on the graph.

_____

2. Calculate the results of the combination of the two linearized equal-percentage characteristics for valve A and valve B based on valve A percent travel when used to split a flow stream. Plot the individual curves for each valve and the total flow based on valve A percent travel.

_____

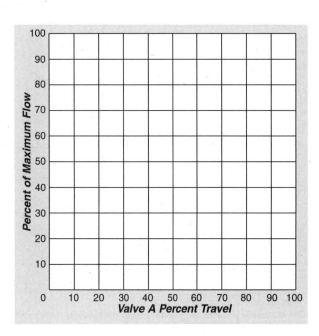

Name _____ Date _____

Activity 46-4—High and Low Switches—Precision Throttling

An electrolyzer is used to generate chlorine and hydrogen gas. The original design called for a control valve to maintain the pressure at setpoint during startup. However, experience has shown that this was not a completely satisfactory control scheme because the valve allowed too much purge gas to pass at the minimum control position. Therefore, it was decided to replace the valve with a precision throttling control valve arrangement containing a large and small control valve, a controller, and a pair of selector relays.

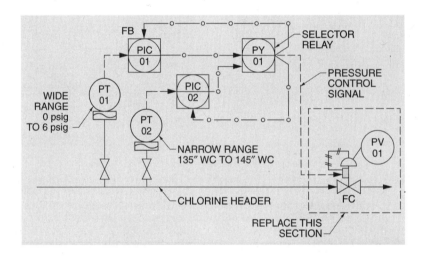

Use the P&ID to answer the following questions.

_____ 1. Use a separate piece of paper to draw a diagram of the required control arrangement to replace the control valve, PV 01.

_____ 2. What setpoint is needed for the high selector relay?

_____ 3. What setpoint is needed for the low selector relay?

_____ 4. If the two control valves are fail closed, what action is required of the controller?

_____ 5. In selecting the large and small control valves, what maximum capacity should the small valve have compared to the large valve?

_____ 6. What flow characteristic should the large valve have?

_____ 7. What flow characteristic should the small valve have?

Name _____ **Date** _____

Activity 46-5—Cascade Control—Distillation Column

A distillation column uses the temperature in the center of the column to control the operation of the column. Steam is used to provide heat to boil the column contents. The steam source has a variable pressure that influences the amount of heat entering the column.

Use the illustration on the back of this page to answer the following questions.

1. What should be cascaded for best control?

_____

_____

_____ **2.** What should be the failure action of the steam control valve, FV 08?

_____ **3.** What is the desired portion of the control signal input to the valve positioner for the steam control valve, FV 08?

_____ **4.** What is the desired action of the column temperature controller?

_____ **5.** What is the desired action of the steam flow controller?

6. Complete the P&ID to include all information from the previous questions.

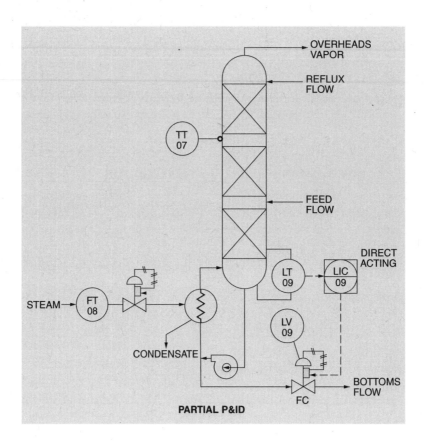

SECTION
12 INSTRUMENTATION AND CONTROL APPLICATIONS

chapter
47
Temperature Control

REVIEW
QUESTIONS

Name _____   **Date** _____

_____ 1. The amount of heat transferred in a heat exchanger depends on the temperature difference, the area of the exchanger, and the ___.
A. convection transfer rate
B. conduction coefficient
C. radiation transfer rate
D. heat transfer coefficient

_____ 2. A thermal expansion ___ must be used to protect heat exchangers from excessive pressure if the fluids can be valved off and trapped in the exchanger.
A. safety relief valve
B. blowout plug
C. heat transfer tube
D. cooling control valve

_____ 3. A ___ is a vertical vessel used in batch processing whose sides and bottom are covered with a steel shell and is used to contain a heating or cooling fluid.
A. cooling tank
B. jacketed reactor
C. distillation column
D. heat extraction vessel

_____ 4. A ___ is any piece of equipment that transfers heat from one material to another.
A. jacketed reactor
B. heat exchanger
C. steam controller
D. refrigeration system

_____ 5. A shell-and-tube heat exchanger typically consists of process fluid contained in ___ that run the length of the shell.
A. tubes
B. plates
C. sheets
D. heads

_____ 6. With a heat exchanger, the temperature of the process fluid can be measured with a temperature-sensing element inserted into the ___ fluid piping.
A. outlet cooling
B. inlet cooling
C. outlet process
D. inlet process

_____ **7.** If the process fluid temperature in a heat exchanger can get high enough to boil the cooling water, a(n) ___ must be used to handle any steam that can be generated.
     A. open vent
     B. relief valve
     C. discharge pump
     D. steam controller

_____ **8.** A steam-controlled heat exchanger has the steam throttled on the ___ to the heat exchanger.
     A. inlet
     B. outlet
     C. relief valve
     D. feed pump

_____ **9.** In almost all steam heating applications, the control valve should have a(n) ___ action.
     A. fail closed
     B. fail open
     C. equal-percentage
     D. linear

_____ **10.** For a steam valve with a fail closed action, the controller action should be ___.
     A. reverse
     B. direct
     C. proportional only
     D. integral only

Name _____ Date _____

Activity 47-1—Cascade Control—Reactor Temperature Control

A batch reactor is used for an exothermic reaction. The temperature in the reactor needs to be raised to the setpoint and kept there until the reaction is complete. While the reaction is proceeding, heat must be removed from the reactor to maintain the temperature setpoint. The amount of cooling required changes over time. The control valves should be split ranged to provide optimum control.

Use the illustration on the back of this page to answer the following questions.

1. What should be cascaded for best control?

_____

_____

_____ 2. What should be the failure action of the chilled water return control valve, TV-1 02?

_____ 3. What should be the failure action of the steam control valve, TV-2 02?

_____ 4. What is the desired portion of the control signal input to the valve positioner for the chilled water return control valve, TV-1 02?

_____ 5. What is the desired portion of the control signal input to the valve positioner for the steam control valve, TV-2 02?

_____ 6. What is the desired action of the reactor temperature controller?

_____ 7. What is the desired action of the circulating water temperature controller?

8. Complete the P&ID to include all the information from the previous questions.

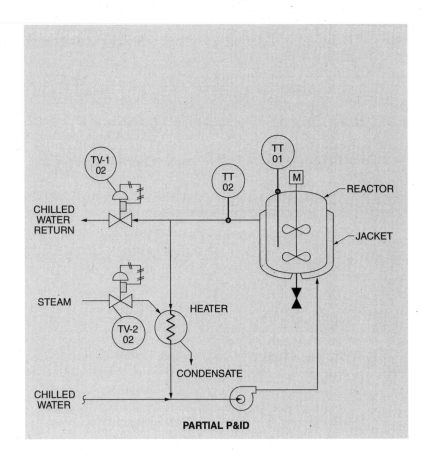

PARTIAL P&ID

SECTION
12 INSTRUMENTATION AND CONTROL APPLICATIONS

chapter
48

Pressure and Level Control

REVIEW
QUESTIONS

Name _____ **Date** _____

_____ **1.** ___ level control is a control arrangement used to start a pump when the level of liquid drops below a setpoint and stop the pump when the liquid level reaches the desired level.
    A. Pump-up
    B. Pump-on
    C. Bridle-activated
    D. Float-balanced

_____ **2.** A ___ is a device that has a stand leg piped to the top and bottom of a boiler or other pressure vessel.
    A. float
    B. bridle
    C. level conductivity probe
    D. level regulator

_____ **3.** Safety in a boiler is provided by a ___, a low level alarm, and a low low level shutdown.
    A. high level alarm
    B. high level shutdown
    C. low level shutdown
    D. all of the above

_____ **4.** A boiler usually uses ___ or float-switch level sensors to determine the water level.
    A. differential pressure
    B. weigh system
    C. radar
    D. conductive-probe

_____ **5.** A boiler feedwater pump can be controlled with two ___.
    A. float-switch level sensors
    B. high-temperature sensors
    C. low-temperature sensors
    D. vacuum pressure gauges

_____ **6.** Conductive-probe level sensors consist of two ___ that pass through a sealed fitting into a stand leg outside a boiler.
    A. impedance tubes
    B. resistors
    C. capacitors
    D. electrodes

_____ **7.** ___ level control is a control arrangement used to control the transfer of collected steam condensate to a common condensate storage tank.

     A. Boiler

     B. Condensate

     C. Pump-down

     D. Pump-up

_____ **8.** When a fail-closed valve action is not compatible with a controller action, a(n) ___ can be used to make the control valve move in the right direction.

     A. equal-percentage action

     B. pressure transmitter

     C. positioner

     D. backpressure system

Name _____   Date _____

Activity 48-1—High and Low Switches—Transfer of Pressure Control

An electrolyzer is used to generate chlorine and hydrogen gas. The pressure of chlorine gas must be controlled to a setpoint of 140 ±0.3″ WC. The setpoint is equivalent to 5.05 psig. At the same time, the differential pressure between the chlorine and hydrogen must be maintained at −4 ±0.75″ WC.

At the startup of the electrolyzer, the chlorine and hydrogen gas generation is too unstable to use the generated gases to bring the gas headers up to operating pressure. Prior to startup, the chlorine header is purged with air and the hydrogen header is purged with nitrogen. The control design uses a wide-range 0 psig to 6 psig pressure transmitter and controller to bring the pressure up to the desired operating point. After reaching the desired operating point, the control automatically switches to a narrow-range 135″ WC to 145″ WC (4.87 psig to 5.24 psig) chlorine pressure transmitter and controller with a setpoint of 140″ WC. The transfer of control is accomplished with a selector relay. Once the headers are up to the operating pressures using the purge gases, the electrolyzer can be energized. The control valve, PV 01, has a fail closed action.

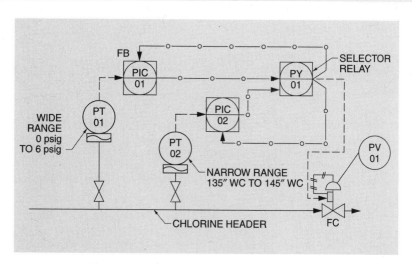

Use the P&ID to answer the following questions.

_____   **1.** Is the selector, PY 01, a high or low selector?

_____   **2.** Does the wide-range controller, PIC 01, use direct or reverse action?

_____   **3.** Does the narrow-range controller, PIC 02, use direct or reverse action?

_____   **4.** What is the setpoint of the wide-range pressure controller so that there will be no interference with the narrow-range pressure controller?

Name _____   Date _____

Activity 48-2—Level Min/Max Control

A variable continuous flow is pumped from a storage tank. The inflow to the tank has a flow capacity greater than the maximum outflow. A precise level control is not needed, but the level should never drop below 20% nor rise above 80% of the maximum tank level. A high-level switch (LSH) is mounted at the 80% elevation and a low-level switch (LSL) is mounted at the 20% elevation. These two switches provide power to a three-way solenoid valve that provides air to a piston actuator on a fail closed valve in the inflow line to the tank.

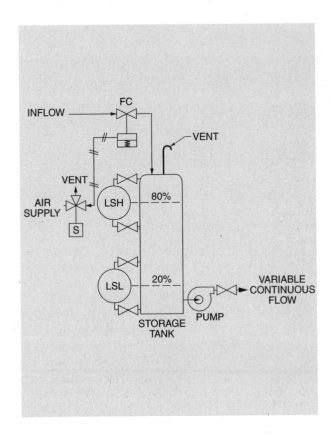

Use the P&ID to answer the following questions.

1. What is the required solenoid valve porting arrangement for the energized and de-energized states?

   _____

   _____

   _____

_____   2. What is the required high-level switch status with no liquid in the tank? (This is the normal state and the one that is shown in a catalog.)

_____   3. What is the required low-level switch status with no liquid in the tank?

4. On a separate piece of paper, draw an electrical ladder diagram that shows how to control the level in the storage tank as described above.

SECTION
12 *INSTRUMENTATION AND CONTROL APPLICATIONS*

chapter
49

Flow Control

REVIEW
QUESTIONS

Name _____   **Date** _____

_____ **1.** A ___ controller is a controller used to control a secondary flow to a predetermined fraction of a primary flow.
    A. continuous
    B. cascade
    C. ratio
    D. multiplier

T   F   **2.** All burner systems need to operate with an excess of combustion air.

_____ **3.** A(n) ___ can be used to fine-tune the air-fuel ratio in a burner system.
    A. conductivity probe
    B. pH meter
    C. flue gas oxygen analyzer
    D. air flow controller

_____ **4.** ___ control is a control strategy used when there is a need to mix two flow streams in a specified ratio.
    A. Continuous flow
    B. Mixing ratio
    C. Multiplier flow
    D. Divider flow

_____ **5.** The overheads flow from a distillation can be divided into two separate constant-ratio flows with ___ control.
    A. splitting ratio
    B. mixing ratio
    C. reflux feedback
    D. overheads feedback

_____ **6.** For a flow ratio controller, the ___ switch is used to determine if the secondary controller is to function as a ratio controller or as a simple stand-alone controller.
    A. track command
    B. remote/local (R/L) setpoint
    C. ratio value
    D. auto/manual (A/M) switch

_____ **7.** In a distillation column, the material with the ___ ends up as overheads vapor while the remaining material ends up as bottoms liquid.

      A. lowest viscosity

      B. highest viscosity

      C. lowest boiling point

      D. highest boiling point

_____ **8.** A lead-lag air-fuel control system is a control system where increases in the flow of combustion air ___ increases in fuel flow.

      A. lead

      B. lag

      C. lead then lag

      D. lag then lead

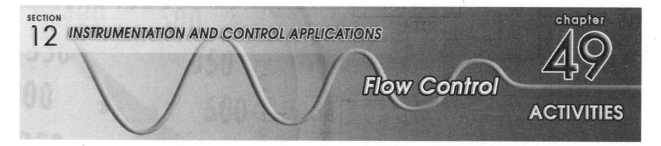
Name _____   Date _____

Activity 49-1—Chlorine Flow Control

A very common form of chlorine vaporizer consists of steam-fed vertical heat exchanger tubes contained in a vertical pressure vessel. The steam is maintained at a constant pressure (which also means a constant temperature).

Liquid chlorine is fed from the dip leg pipe of a chlorine rail car. During warm weather, the vapor pressure of the chlorine in the tank car is usually sufficient to provide the pressure needed to carry the vaporized chlorine to its end use. In cool weather, the chlorine tank car is kept under pressure with a pad of dry air or nitrogen. Wet chlorine is very corrosive and operating conditions are designed to prevent water from getting into the chlorine.

A general equation for heat transfer is as follows:

$$\Delta H = U \times A \times (T_1 - T_2)$$

where

ΔH = heat transferred, in Btu
U = fixed heat transfer constant, in Btu/sq ft/°F
A = heat transfer area, in sq ft
T_1 = higher temperature, in °F
T_2 = lower temperature, in °F

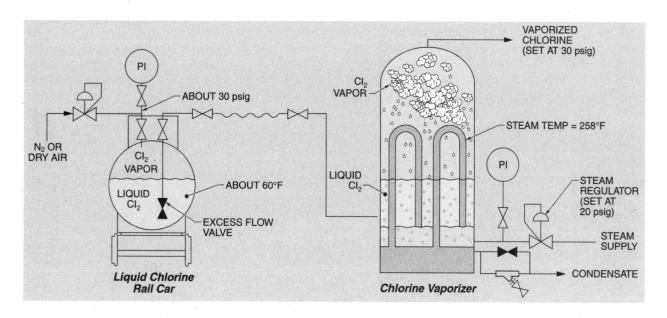

Liquid Chlorine Rail Car

Chlorine Vaporizer

Use the illustration on the previous page to answer the following questions.

1. Explain how the vaporizer works to vary the chlorine evaporation rate to match the vapor demand.

_____

_____

2. What happens in the vaporizer if the downstream process stops using chlorine?

_____

_____

3. What happens in the vaporizer if the downstream process requires more chlorine vapor than the vaporizer can supply?

_____

_____

4. What safety measures can be added to prevent hazardous carryover of liquid chlorine from the vaporizer to the downstream process?

_____

_____

5. Why is the steam pressure selected to be lower than the chlorine pressure?

_____

_____

Name _____   Date _____

Activity 49-2—Lead-Lag Air-Fuel Ratio

A natural gas fired boiler uses a digital lead-lag air-fuel ratio control system with oxygen trim. The oxygen trim system is limited to a maximum change of the air or fuel flow of ±10% of the maximum air or fuel flows. The maximum flow rate for natural gas is 750 scfh and for air is 7500 scfh. The maximum steam pressure is 150 psig and the oxygen analyzer range is 0% to 5%.

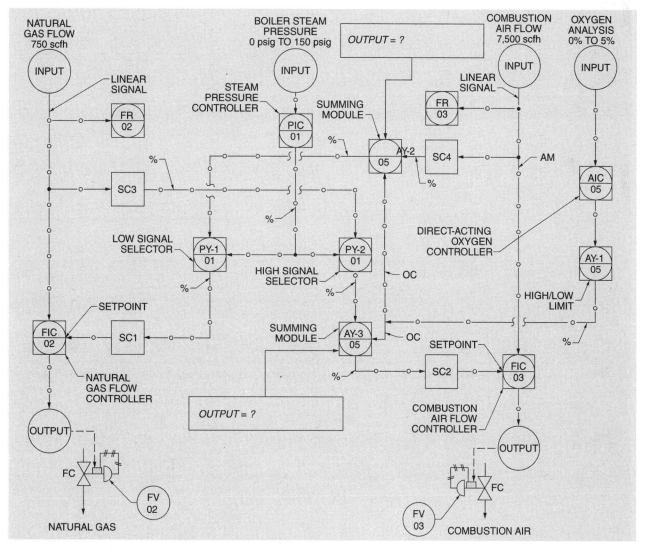

Use the control arrangement shown in the figure to answer the following questions.

1. What is the scaling factor for SC1?

   _____

   _____

2. What is the scaling factor for SC2?

   _____

   _____

3. What is the scaling factor for SC3?

   _____

   _____

4. What is the scaling factor for SC4?

   _____

   _____

5. What is the equation for the summing module AY-2?

   _____

   _____

6. What is the equation for the summing module AY-3?

   _____

   _____

Name _____ Date _____

Activity 49-3—Flow Ratio Control Scales

There are a number of different types of equipment that can be used to implement flow ratio control. These include digital, electronic, and pneumatic controllers. Some of these controllers use the ratio of the transmission signals and some use the actual flow engineering units.

Row	Type of Controller	Primary Flow Range	Secondary Flow Range	Primary Flow	Secondary Flow
1	Digital controller, linear transmitters	150 gpm	50 gpm	108 gpm	35 gpm
2	Pneumatic controller, linear transmitters	75 gpm	15 gpm	52 gpm	10.8 gpm
3	Electric controller, linear transmitters	35 gpm	50 gpm	25 gpm	33 gpm
4	Digital controller, linear transmitters	1000 scfh	750 scfh	512 scfh	400 scfh

Use the table to answer the following questions.

_____ **1.** What is the flow ratio setting for row 1 of the table?

_____ **2.** What is the flow ratio setting for row 2 of the table?

_____ **3.** What is the flow ratio setting for row 3 of the table?

_____ **4.** What is the flow ratio setting for row 4 of the table?

Name _____ Date _____

Activity 49-4—Flow Ratio Control Diagrams

Flow ratio control can be shown on piping and instrumentation diagrams. This activity is a situation where the flow to "B" is to be a fixed ratio of the total flow except that the flow to "B" is to be kept at or above a specified minimum flow. The level control valve, LV, is controlled by a feed tank level control system.

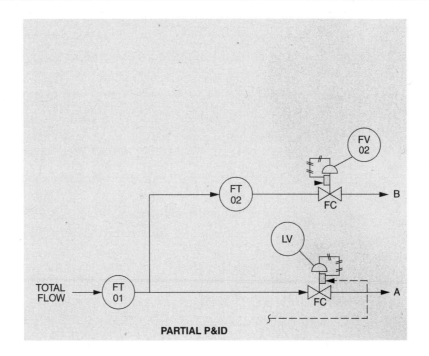

PARTIAL P&ID

Use the P&ID to answer the following questions.

_____ **1.** What is the primary flow stream?

_____ **2.** What is the secondary flow stream?

3. Use standard ISA symbols to complete the P&ID. Show any required controllers, selectors, control signals, and feedback signals required to implement the required control system.

SECTION
12 INSTRUMENTATION AND CONTROL APPLICATIONS

chapter
50

Analysis and Multivariable Control

REVIEW
QUESTIONS

Name _____   Date _____

_____ **1.** ___ is formed when steam condenses.
- A. Entropy
- B. Superheated steam
- C. Condensate
- D. Enthalpy

_____ **2.** Conveyors are commonly used to move ___ from one location to another.
- A. dry materials
- B. liquids
- C. vapors
- D. gases

_____ **3.** The quality of returned condensate can be measured with a ___ instrument.
- A. conductivity
- B. viscosity
- C. refractive index
- D. density

_____ **4.** A control system that maintains constant flow out of a tank, except when a tank level is at a level limit, consists of a standard flow control loop and a ___ level controller.
- A. cascade
- B. gap action
- C. ratio
- D. setpoint

_____ **5.** A ___ is a boiler in which water passes through tubes surrounded by gases of combustion.
- A. boiler drum
- B. condensate collector tube
- C. firetube boiler
- D. watertube boiler

_____ **6.** A sudden increase in steam demand temporarily lowers boiler steam pressure and causes the water level in the steam drum to ___ because bubbles of steam suddenly form in the water.
- A. double
- B. decrease
- C. shrink
- D. swell

_____   **7.** A boiler control system can combine a drum level measurement with a ___ measurement to improve control of the water level in the boiler.
  A. steam flow
  B. water pressure
  C. steam temperature
  D. condensate conductivity

_____   **8.** A boiler drum level controller needs to be ___ with proportional and integral functions.
  A. direct-acting
  B. reverse-acting
  C. quick-opening
  D. equal-percentage

Name _____ Date _____

Activity 50-1—High and Low Switches—Flow and Pressure Control

A particularly gaseous material needs to be fed to a process at a designated controlled flow rate as long as the pressure of the process does not drop below a specified value. A downstream pressure controller is arranged with the flow controller to always maintain the specified pressure even if the flow is greater than its setpoint. The control valve has a fail open action.

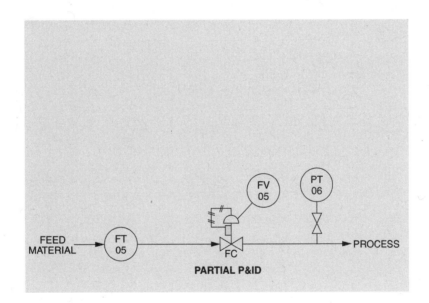

PARTIAL P&ID

1. Complete the control diagram so that its design will satisfy the stated requirements.

_____ 2. What type of selector is needed?

_____ 3. What is the proper action of the flow controller?

_____ 4. What is the proper action of the pressure controller?

5. Why are the feedback signals needed?

_____

_____

_____

Name _____   Date _____

Chlorine Vaporizer Control

A large-capacity chlorine vaporizer is shown in the illustration. The rate of vaporization of the chlorine is directly related to the amount of heat energy being supplied from the steam. The steam temperature and pressure in the tubes are dependent on the heat transfer constant of the steam tubes.

   The level of the liquid chlorine is controlled with a feed valve and level control system to keep the liquid chlorine above the tops of the steam tubes. The pressure of the vaporized chlorine is measured and used to control the steam flow into the vaporizer.

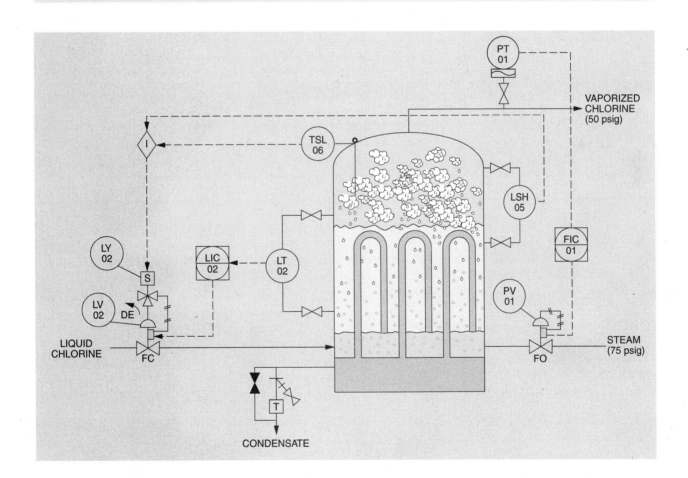

An understanding of the relationships among temperature, pressure, and heat capacity of liquids and vapors at their boiling points is required to answer the following questions.

1. Why is this vaporizer system able to provide a very steady pressure control? Consider what happens to the boiling chlorine with increases and decreases in vapor pressure.

   _____

   _____

   _____

   _____

2. What problems can be encountered at low vaporization rates?

   _____

   _____

   _____

   _____

3. What is the purpose of the TSL-06 and LSH-05 switches?

   _____

   _____

   _____

   _____

Pt F

100Ω Platinum RTD — *0.00385 coefficient*
temperature in °F

°F	0	1	2	3	4	5	6	7	8	9	10	°F
						Resistance in Ohms						
-320	20.44	20.20	19.96	19.72	19.48	19.24	19.00	18.76	18.52			-320
-310	22.83	22.59	22.35	22.11	21.87	21.63	21.39	21.16	20.92	20.68	20.44	-310
-300	25.20	24.97	24.73	24.49	24.25	24.02	23.78	23.54	23.30	23.06	22.83	-300
-290	27.57	27.33	27.10	26.86	26.62	26.39	26.15	25.91	25.68	25.44	25.20	-290
-280	29.93	29.69	29.46	29.22	28.98	28.75	28.51	28.28	28.04	27.81	27.57	-280
-270	32.27	32.04	31.80	31.57	31.34	31.10	30.87	30.63	30.40	30.16	29.93	-270
-260	34.61	34.38	34.14	33.91	33.68	33.44	33.21	32.98	32.74	32.51	32.27	-260
-250	36.94	36.71	36.47	36.24	36.01	35.78	35.54	35.31	35.08	34.84	34.61	-250
-240	39.26	39.03	38.80	38.56	38.33	38.10	37.87	37.64	37.40	37.17	36.94	-240
-230	41.57	41.34	41.11	40.88	40.65	40.42	40.19	39.95	39.72	39.49	39.26	-230
-220	43.88	43.65	43.42	43.19	42.96	42.73	42.49	42.26	42.03	41.80	41.57	-220
-210	46.17	45.94	45.71	45.48	45.26	45.03	44.80	44.57	44.34	44.11	43.88	-210
-200	48.46	48.23	48.00	47.78	47.55	47.32	47.09	46.86	46.63	46.40	46.17	-200
-190	50.74	50.52	50.29	50.06	49.83	49.60	49.38	49.15	48.92	48.69	48.46	-190
-180	53.02	52.79	52.56	52.34	52.11	51.88	51.65	51.43	51.20	50.97	50.74	-180
-170	55.29	55.06	54.83	54.61	54.38	54.15	53.93	53.70	53.47	53.25	53.02	-170
-160	57.55	57.32	57.10	56.87	56.65	56.42	56.19	55.97	55.74	55.51	55.29	-160
-150	59.81	59.58	59.35	59.13	58.90	58.68	58.45	58.23	58.00	57.78	57.55	-150
-140	62.06	61.83	61.61	61.38	61.16	60.93	60.71	60.48	60.26	60.03	59.81	-140
-130	64.30	64.08	63.85	63.63	63.40	63.18	62.95	62.73	62.50	62.28	62.06	-130
-120	66.54	66.31	66.09	65.87	65.64	65.42	65.20	64.97	64.75	64.52	64.30	-120
-110	68.77	68.55	68.33	68.10	67.88	67.66	67.43	67.21	66.99	66.76	66.54	-110
-100	71.00	70.78	70.55	70.33	70.11	69.89	69.66	69.44	69.22	68.99	68.77	-100
-90	73.22	73.00	72.78	72.56	72.33	72.11	71.89	71.67	71.45	71.22	71.00	-90
-80	75.44	75.22	75.00	74.78	74.55	74.33	74.11	73.89	73.67	73.45	73.22	-80
-70	77.66	77.43	77.21	76.99	76.77	76.55	76.33	76.11	75.88	75.66	75.44	-70
-60	79.86	79.64	79.42	79.20	78.98	78.76	78.54	78.32	78.10	77.88	77.66	-60
-50	82.07	81.85	81.63	81.41	81.19	80.97	80.75	80.53	80.31	80.09	79.86	-50
-40	84.27	84.05	83.83	83.61	83.39	83.17	82.95	82.73	82.51	82.29	82.07	-40
-30	86.47	86.25	86.03	85.81	85.59	85.37	85.15	84.93	84.71	84.49	84.27	-30
-20	88.66	88.44	88.22	88.00	87.78	87.56	87.34	87.13	86.91	86.69	86.47	-20
-10	90.85	90.63	90.41	90.19	89.97	89.75	89.54	89.32	89.10	88.88	88.66	-10
0	93.03	92.82	92.60	92.38	92.16	91.94	91.72	91.50	91.29	91.07	90.85	0
0	93.03	93.25	93.47	93.69	93.91	94.12	94.34	94.56	94.78	95.00	95.21	0
10	95.21	95.43	95.65	95.87	96.09	96.30	96.52	96.74	96.96	97.17	97.39	10
20	97.39	97.61	97.83	98.04	98.26	98.48	98.70	98.91	99.13	99.35	99.57	20
30	99.57	99.78	100.00	100.22	100.43	100.65	100.87	101.09	101.30	101.52	101.74	30
40	101.74	101.95	102.17	102.39	102.60	102.82	103.04	103.25	103.47	103.69	103.90	40
50	103.90	104.12	104.34	104.55	104.77	104.98	105.20	105.42	105.63	105.85	106.07	50
60	106.07	106.28	106.50	106.71	106.93	107.15	107.36	107.58	107.79	108.01	108.23	60
70	108.23	108.44	108.66	108.87	109.09	109.30	109.52	109.73	109.95	110.17	110.38	70
80	110.38	110.60	110.81	111.03	111.24	111.46	111.67	111.89	112.10	112.32	112.53	80
90	112.53	112.75	112.96	113.18	113.39	113.61	113.82	114.04	114.25	114.47	114.68	90
100	114.68	114.90	115.11	115.33	115.54	115.76	115.97	116.18	116.40	116.61	116.83	100
110	116.83	117.04	117.26	117.47	117.68	117.90	118.11	118.33	118.54	118.76	118.97	110
120	118.97	119.18	119.40	119.61	119.82	120.04	120.25	120.47	120.68	120.89	121.11	120
130	121.11	121.32	121.53	121.75	121.96	122.18	122.39	122.60	122.82	123.03	123.24	130
140	123.24	123.46	123.67	123.88	124.09	124.31	124.52	124.73	124.95	125.16	125.37	140
°F	0	1	2	3	4	5	6	7	8	9	10	°F

Pyro MATION, INC.

F

100Ω *Platinum RTD* — *0.00385 coefficient*
temperature in °*F*

°F	0	1	2	3	4	5	6	7	8	9	10	°F
						Resistance in Ohms						
150	125.37	125.59	125.80	126.01	126.22	126.44	126.65	126.86	127.08	127.29	127.50	150
160	127.50	127.71	127.93	128.14	128.35	128.56	128.78	128.99	129.20	129.41	129.62	160
170	129.62	129.84	130.05	130.26	130.47	130.68	130.90	131.11	131.32	131.53	131.74	170
180	131.74	131.96	132.17	132.38	132.59	132.80	133.01	133.23	133.44	133.65	133.86	180
190	133.86	134.07	134.28	134.50	134.71	134.92	135.13	135.34	135.55	135.76	135.97	190
200	135.97	136.19	136.40	136.61	136.82	137.03	137.24	137.45	137.66	137.87	138.08	200
210	138.08	138.29	138.51	138.72	138.93	139.14	139.35	139.56	139.77	139.98	140.19	210
220	140.19	140.40	140.61	140.82	141.03	141.24	141.45	141.66	141.87	142.08	142.29	220
230	142.29	142.50	142.71	142.92	143.13	143.34	143.55	143.76	143.97	144.18	144.39	230
240	144.39	144.60	144.81	145.02	145.23	145.44	145.65	145.86	146.07	146.28	146.49	240
250	146.49	146.70	146.91	147.11	147.32	147.53	147.74	147.95	148.16	148.37	148.58	250
260	148.58	148.79	149.00	149.21	149.41	149.62	149.83	150.04	150.25	150.46	150.67	260
270	150.67	150.88	151.08	151.29	151.50	151.71	151.92	152.13	152.33	152.54	152.75	270
280	152.75	152.96	153.17	153.38	153.58	153.79	154.00	154.21	154.42	154.62	154.83	280
290	154.83	155.04	155.25	155.46	155.66	155.87	156.08	156.29	156.49	156.70	156.91	290
300	156.91	157.12	157.33	157.53	157.74	157.95	158.15	158.36	158.57	158.78	158.98	300
310	158.98	159.19	159.40	159.61	159.81	160.02	160.23	160.43	160.64	160.85	161.05	310
320	161.05	161.26	161.47	161.67	161.88	162.09	162.29	162.50	162.71	162.91	163.12	320
330	163.12	163.33	163.53	163.74	163.95	164.15	164.36	164.57	164.77	164.98	165.18	330
340	165.18	165.39	165.60	165.80	166.01	166.21	166.42	166.63	166.83	167.04	167.24	340
350	167.24	167.45	167.66	167.86	168.07	168.27	168.48	168.68	168.89	169.09	169.30	350
360	169.30	169.51	169.71	169.92	170.12	170.33	170.53	170.74	170.94	171.15	171.35	360
370	171.35	171.56	171.76	171.97	172.17	172.38	172.58	172.79	172.99	173.20	173.40	370
380	173.40	173.61	173.81	174.02	174.22	174.43	174.63	174.83	175.04	175.24	175.45	380
390	175.45	175.65	175.86	176.06	176.26	176.47	176.67	176.88	177.08	177.29	177.49	390
400	177.49	177.69	177.90	178.10	178.30	178.51	178.71	178.92	179.12	179.32	179.53	400
410	179.53	179.73	179.93	180.14	180.34	180.55	180.75	180.95	181.16	181.36	181.56	410
420	181.56	181.77	181.97	182.17	182.38	182.58	182.78	182.98	183.19	183.39	183.59	420
430	183.59	183.80	184.00	184.20	184.40	184.61	184.81	185.01	185.22	185.42	185.62	430
440	185.62	185.82	186.03	186.23	186.43	186.63	186.84	187.04	187.24	187.44	187.65	440
450	187.65	187.85	188.05	188.25	188.45	188.66	188.86	189.06	189.26	189.46	189.67	450
460	189.67	189.87	190.07	190.27	190.47	190.67	190.88	191.08	191.28	191.48	191.68	460
470	191.68	191.88	192.09	192.29	192.49	192.69	192.89	193.09	193.29	193.49	193.70	470
480	193.70	193.90	194.10	194.30	194.50	194.70	194.90	195.10	195.30	195.50	195.71	480
490	195.71	195.91	196.11	196.31	196.51	196.71	196.91	197.11	197.31	197.51	197.71	490
500	197.71	197.91	198.11	198.31	198.51	198.71	198.91	199.11	199.31	199.51	199.71	500
510	199.71	199.91	200.11	200.31	200.51	200.71	200.91	201.11	201.31	201.51	201.71	510
520	201.71	201.91	202.11	202.31	202.51	202.71	202.91	203.11	203.31	203.51	203.71	520
530	203.71	203.91	204.11	204.31	204.51	204.71	204.90	205.10	205.30	205.50	205.70	530
540	205.70	205.90	206.10	206.30	206.50	206.70	206.89	207.09	207.29	207.49	207.69	540
550	207.69	207.89	208.09	208.29	208.48	208.68	208.88	209.08	209.28	209.48	209.67	550
560	209.67	209.87	210.07	210.27	210.47	210.67	210.86	211.06	211.26	211.46	211.66	560
570	211.66	211.85	212.05	212.25	212.45	212.64	212.84	213.04	213.24	213.44	213.63	570
580	213.63	213.83	214.03	214.23	214.42	214.62	214.82	215.02	215.21	215.41	215.61	580
590	215.61	215.80	216.00	216.20	216.40	216.59	216.79	216.99	217.18	217.38	217.58	590
600	217.58	217.77	217.97	218.17	218.37	218.56	218.76	218.96	219.15	219.35	219.55	600
610	219.55	219.74	219.94	220.13	220.33	220.53	220.72	220.92	221.12	221.31	221.51	610
620	221.51	221.70	221.90	222.10	222.29	222.49	222.68	222.88	223.08	223.27	223.47	620
630	223.47	223.66	223.86	224.06	224.25	224.45	224.64	224.84	225.03	225.23	225.42	630
640	225.42	225.62	225.82	226.01	226.21	226.40	226.60	226.79	226.99	227.18	227.38	640
°F	0	1	2	3	4	5	6	7	8	9	10	°F

Pyro **MATION, INC.**

ITS-90 Table for Type J Thermocouple...

Thermoelectric Voltage in mV

°C	0	−1	−2	−3	−4	−5	−6	−7	−8	−9	−10
−210	−8.095										
−200	−7.890	−7.912	−7.934	−7.955	−7.976	−7.996	−8.017	−8.037	−8.057	−8.076	−8.095
−190	−7.659	−7.683	−7.707	−7.731	−7.755	−7.778	−7.801	−7.824	−7.846	−7.868	−7.890
−180	−7.403	−7.429	−7.456	−7.482	−7.508	−7.534	−7.559	−7.585	−7.610	−7.634	−7.659
−170	−7.123	−7.152	−7.181	−7.209	−7.237	−7.265	−7.293	−7.321	−7.348	−7.376	−7.403
−160	−6.821	−6.853	−6.883	−6.914	−6.944	−6.975	−7.005	−7.035	−7.064	−7.094	−7.123
−150	−6.500	−6.533	−6.566	−6.598	−6.631	−6.663	−6.695	−6.727	−6.759	−6.790	−6.821
−140	−6.159	−6.194	−6.229	−6.263	−6.298	−6.332	−6.366	−6.400	−6.433	−6.467	−6.500
−130	−5.801	−5.838	−5.874	−5.910	−5.946	−5.982	−6.018	−6.054	−6.089	−6.124	−6.159
−120	−5.426	−5.465	−5.503	−5.541	−5.578	−5.616	−5.653	−5.690	−5.727	−5.764	−5.801
−110	−5.037	−5.076	−5.116	−5.155	−5.194	−5.233	−5.272	−5.311	−5.350	−5.388	−5.426
−100	−4.633	−4.674	−4.714	−4.755	−4.796	−4.836	−4.877	−4.917	−4.957	−4.997	−5.037
−90	−4.215	−4.257	−4.300	−4.342	−4.384	−4.425	−4.467	−4.509	−4.550	−4.591	−4.633
−80	−3.786	−3.829	−3.872	−3.916	−3.959	−4.002	−4.045	−4.088	−4.130	−4.173	−4.215
−70	−3.344	−3.389	−3.434	−3.478	−3.522	−3.566	−3.610	−3.654	−3.698	−3.742	−3.786
−60	−2.893	−2.938	−2.984	−3.029	−3.075	−3.120	−3.165	−3.210	−3.255	−3.300	−3.344
−50	−2.431	−2.478	−2.524	−2.571	−2.617	−2.663	−2.709	−2.755	−2.801	−2.847	−2.893
−40	−1.961	−2.008	−2.055	−2.103	−2.150	−2.197	−2.244	−2.291	−2.338	−2.385	−2.431
−30	−1.482	−1.530	−1.578	−1.626	−1.674	−1.722	−1.770	−1.818	−1.865	−1.913	−1.961
−20	−0.995	−1.044	−1.093	−1.142	−1.190	−1.239	−1.288	−1.336	−1.385	−1.433	−1.482
−10	−0.501	−0.550	−0.600	−0.650	−0.699	−0.749	−0.798	−0.847	−0.896	−0.946	−0.995
0	0.000	−0.050	−0.101	−0.151	−0.201	−0.251	−0.301	−0.351	−0.401	−0.451	−0.501

°C	0	1	2	3	4	5	6	7	8	9	10
0	0.000	0.050	0.101	0.151	0.202	0.253	0.303	0.354	0.405	0.456	0.507
10	0.507	0.558	0.609	0.660	0.711	0.762	0.814	0.865	0.916	0.968	1.019
20	1.019	1.071	1.122	1.174	1.226	1.277	1.329	1.381	1.433	1.485	1.537
30	1.537	1.589	1.641	1.693	1.745	1.797	1.849	1.902	1.954	2.006	2.059
40	2.059	2.111	2.164	2.216	2.269	2.322	2.374	2.427	2.480	2.532	2.585
50	2.585	2.638	2.691	2.744	2.797	2.850	2.903	2.956	3.009	3.062	3.116
60	3.116	3.169	3.222	3.275	3.329	3.382	3.436	3.489	3.543	3.596	3.650
70	3.650	3.703	3.757	3.810	3.864	3.918	3.971	4.025	4.079	4.133	4.187
80	4.187	4.240	4.294	4.348	4.402	4.456	4.510	4.564	4.618	4.672	4.726
90	4.726	4.781	4.835	4.889	4.943	4.997	5.052	5.106	5.160	5.215	5.269

...ITS-90 Table for Type J Thermocouple...

Thermoelectric Voltage in mV

°C	0	1	2	3	4	5	6	7	8	9	10
100	5.269	5.323	5.378	5.432	5.487	5.541	5.595	5.650	5.705	5.759	5.814
110	5.814	5.868	5.923	5.977	6.032	6.087	6.141	6.196	6.251	6.306	6.360
120	6.360	6.415	6.470	6.525	6.579	6.634	6.689	6.744	6.799	6.854	6.909
130	6.909	6.964	7.019	7.074	7.129	7.184	7.239	7.294	7.349	7.404	7.459
140	7.459	7.514	7.569	7.624	7.679	7.734	7.789	7.844	7.900	7.955	8.010
150	8.010	8.065	8.120	8.175	8.231	8.286	8.341	8.396	8.452	8.507	8.562
160	8.562	8.618	8.673	8.728	8.783	8.839	8.894	8.949	9.005	9.060	9.115
170	9.115	9.171	9.226	9.282	9.337	9.392	9.448	9.503	9.559	9.614	9.669
180	9.669	9.725	9.780	9.836	9.891	9.947	10.002	10.057	10.113	10.168	10.224
190	10.224	10.279	10.335	10.390	10.446	10.501	10.557	10.612	10.668	10.723	10.779
200	10.779	10.834	10.890	10.945	11.001	11.056	11.112	11.167	11.223	11.278	11.334
210	11.334	11.389	11.445	11.501	11.556	11.612	11.667	11.723	11.778	11.834	11.889
220	11.889	11.945	12.000	12.056	12.111	12.167	12.222	12.278	12.334	12.389	12.445
230	12.445	12.500	12.556	12.611	12.667	12.722	12.778	12.833	12.889	12.944	13.000
240	13.000	13.056	13.111	13.167	13.222	13.278	13.333	13.389	13.444	13.500	13.555
250	13.555	13.611	13.666	13.722	13.777	13.833	13.888	13.944	13.999	14.055	14.110
260	14.110	14.166	14.221	14.277	14.332	14.388	14.443	14.499	14.554	14.609	14.665
270	14.665	14.720	14.776	14.831	14.887	14.942	14.998	15.053	15.109	15.164	15.219
280	15.219	15.275	15.330	15.386	15.441	15.496	15.552	15.607	15.663	15.718	15.773
290	15.773	15.829	15.884	15.940	15.995	16.050	16.106	16.161	16.216	16.272	16.327
300	16.327	16.383	16.438	16.493	16.549	16.604	16.659	16.715	16.770	16.825	16.881
310	16.881	16.936	16.991	17.046	17.102	17.157	17.212	17.268	17.323	17.378	17.434
320	17.434	17.489	17.544	17.599	17.655	17.710	17.765	17.820	17.876	17.931	17.986
330	17.986	18.041	18.097	18.152	18.207	18.262	18.318	18.373	18.428	18.483	18.538
340	18.538	18.594	18.649	18.704	18.759	18.814	18.870	18.925	18.980	19.035	19.090
350	19.090	19.146	19.201	19.256	19.311	19.366	19.422	19.477	19.532	19.587	19.642
360	19.642	19.697	19.753	19.808	19.863	19.918	19.973	20.028	20.083	20.139	20.194
370	20.194	20.249	20.304	20.359	20.414	20.469	20.525	20.580	20.635	20.690	20.745
380	20.745	20.800	20.855	20.911	20.966	21.021	21.076	21.131	21.186	21.241	21.297
390	21.297	21.352	21.407	21.462	21.517	21.572	21.627	21.683	21.738	21.793	21.848
400	21.848	21.903	21.958	22.014	22.069	22.124	22.179	22.234	22.289	22.345	22.400
410	22.400	22.455	22.510	22.565	22.620	22.676	22.731	22.786	22.841	22.896	22.952
420	22.952	23.007	23.062	23.117	23.172	23.228	23.283	23.338	23.393	23.449	23.504
430	23.504	23.559	23.614	23.670	23.725	23.780	23.835	23.891	23.946	24.001	24.057
440	24.057	24.112	24.167	24.223	24.278	24.333	24.389	24.444	24.499	24.555	24.610

...ITS-90 Table for Type J Thermocouple...

Thermoelectric Voltage in mV

°C	0	1	2	3	4	5	6	7	8	9	10
450	24.610	24.665	24.721	24.776	24.832	24.887	24.943	24.998	25.053	25.109	25.164
460	25.164	25.220	25.275	25.331	25.386	25.442	25.497	25.553	25.608	25.664	25.720
470	25.720	25.775	25.831	25.886	25.942	25.998	26.053	26.109	26.165	26.220	26.276
480	26.276	26.332	26.387	26.443	26.499	26.555	26.610	26.666	26.722	26.778	26.834
490	26.834	26.889	26.945	27.001	27.057	27.113	27.169	27.225	27.281	27.337	27.393
500	27.393	27.449	27.505	27.561	27.617	27.673	27.729	27.785	27.841	27.897	27.953
510	27.953	28.010	28.066	28.122	28.178	28.234	28.291	28.347	28.403	28.460	28.516
520	28.516	28.572	28.629	28.685	28.741	28.798	28.854	28.911	28.967	29.024	29.080
530	29.080	29.137	29.194	29.250	29.307	29.363	29.420	29.477	29.534	29.590	29.647
540	29.647	29.704	29.761	29.818	29.874	29.931	29.988	30.045	30.102	30.159	30.216
550	30.216	30.273	30.330	30.387	30.444	30.502	30.559	30.616	30.673	30.730	30.788
560	30.788	30.845	30.902	30.960	31.017	31.074	31.132	31.189	31.247	31.304	31.362
570	31.362	31.419	31.477	31.535	31.592	31.650	31.708	31.766	31.823	31.881	31.939
580	31.939	31.997	32.055	32.113	32.171	32.229	32.287	32.345	32.403	32.461	32.519
590	32.519	32.577	32.636	32.694	32.752	32.810	32.869	32.927	32.985	33.044	33.102
600	33.102	33.161	33.219	33.278	33.337	33.395	33.454	33.513	33.571	33.630	33.689
610	33.689	33.748	33.807	33.866	33.925	33.984	34.043	34.102	34.161	34.220	34.279
620	34.279	34.338	34.397	34.457	34.516	34.575	34.635	34.694	34.754	34.813	34.873
630	34.873	34.932	34.992	35.051	35.111	35.171	35.230	35.290	35.350	35.410	35.470
640	35.470	35.530	35.590	35.650	35.710	35.770	35.830	35.890	35.950	36.010	36.071
650	36.071	36.131	36.191	36.252	36.312	36.373	36.433	36.494	36.554	36.615	36.675
660	36.675	36.736	36.797	36.858	36.918	36.979	37.040	37.101	37.162	37.223	37.284
670	37.284	37.345	37.406	37.467	37.528	37.590	37.651	37.712	37.773	37.835	37.896
680	37.896	37.958	38.019	38.081	38.142	38.204	38.265	38.327	38.389	38.450	38.512
690	38.512	38.574	38.636	38.698	38.760	38.822	38.884	38.946	39.008	39.070	39.132
700	39.132	39.194	39.256	39.318	39.381	39.443	39.505	39.568	39.630	39.693	39.755
710	39.755	39.818	39.880	39.943	40.005	40.068	40.131	40.193	40.256	40.319	40.382
720	40.382	40.445	40.508	40.570	40.633	40.696	40.759	40.822	40.886	40.949	41.012
730	41.012	41.075	41.138	41.201	41.265	41.328	41.391	41.455	41.518	41.581	41.645
740	41.645	41.708	41.772	41.835	41.899	41.962	42.026	42.090	42.153	42.217	42.281
750	42.281	42.344	42.408	42.472	42.536	42.599	42.663	42.727	42.791	42.855	42.919
760	42.919	42.983	43.047	43.111	43.175	43.239	43.303	43.367	43.431	43.495	43.559
770	43.559	43.624	43.688	43.752	43.817	43.881	43.945	44.010	44.074	44.139	44.203
780	44.203	44.267	44.332	44.396	44.461	44.525	44.590	44.655	44.719	44.784	44.848
790	44.848	44.913	44.977	45.042	45.107	45.171	45.236	45.301	45.365	45.430	45.494

...ITS-90 Table for Type J Thermocouple...

Thermoelectric Voltage in mV

°C	0	1	2	3	4	5	6	7	8	9	10
800	45.494	45.559	45.624	45.688	45.753	45.818	45.882	45.947	46.011	46.076	46.141
810	46.141	46.205	46.270	46.334	46.399	46.464	46.528	46.593	46.657	46.722	46.786
820	46.786	46.851	46.915	46.980	47.044	47.109	47.173	47.238	47.302	47.367	47.431
830	47.431	47.495	47.560	47.624	47.688	47.753	47.817	47.881	47.946	48.010	48.074
840	48.074	48.138	48.202	48.267	48.331	48.395	48.459	48.523	48.587	48.651	48.715
850	48.715	48.779	48.843	48.907	48.971	49.034	49.098	49.162	49.226	49.290	49.353
860	49.353	49.417	49.481	49.544	49.608	49.672	49.735	49.799	49.862	49.926	49.989
870	49.989	50.052	50.116	50.179	50.243	50.306	50.369	50.432	50.495	50.559	50.622
880	50.622	50.685	50.748	50.811	50.874	50.937	51.000	51.063	51.126	51.188	51.251
890	51.251	51.314	51.377	51.439	51.502	51.565	51.627	51.690	51.752	51.815	51.877
900	51.877	51.940	52.002	52.064	52.127	52.189	52.251	52.314	52.376	52.438	52.500
910	52.500	52.562	52.624	52.686	52.748	52.810	52.872	52.934	52.996	53.057	53.119
920	53.119	53.181	53.243	53.304	53.366	53.427	53.489	53.550	53.612	53.673	53.735
930	53.735	53.796	53.857	53.919	53.980	54.041	54.102	54.164	54.225	54.286	54.347
940	54.347	54.408	54.469	54.530	54.591	54.652	54.713	54.773	54.834	54.895	54.956
950	54.956	55.016	55.077	55.138	55.198	55.259	55.319	55.380	55.440	55.501	55.561
960	55.561	55.622	55.682	55.742	55.803	55.863	55.923	55.983	56.043	56.104	56.164
970	56.164	56.224	56.284	56.344	56.404	56.464	56.524	56.584	56.643	56.703	56.763
980	56.763	56.823	56.883	56.942	57.002	57.062	57.121	57.181	57.240	57.300	57.360
990	57.360	57.419	57.479	57.538	57.597	57.657	57.716	57.776	57.835	57.894	57.953
1000	57.953	58.013	58.072	58.131	58.190	58.249	58.309	58.368	58.427	58.486	58.545
1010	58.545	58.604	58.663	58.722	58.781	58.840	58.899	58.957	59.016	59.075	59.134
1020	59.134	59.193	59.252	59.310	59.369	59.428	59.487	59.545	59.604	59.663	59.721
1030	59.721	59.780	59.838	59.897	59.956	60.014	60.073	60.131	60.190	60.248	60.307
1040	60.307	60.365	60.423	60.482	60.540	60.599	60.657	60.715	60.774	60.832	60.890
1050	60.890	60.949	61.007	61.065	61.123	61.182	61.240	61.298	61.356	61.415	61.473
1060	61.473	61.531	61.589	61.647	61.705	61.763	61.822	61.880	61.938	61.996	62.054
1070	62.054	62.112	62.170	62.228	62.286	62.344	62.402	62.460	62.518	62.576	62.634
1080	62.634	62.692	62.750	62.808	62.866	62.924	62.982	63.040	63.098	63.156	63.214
1090	63.214	63.271	63.329	63.387	63.445	63.503	63.561	63.619	63.677	63.734	63.792
1100	63.792	63.850	63.908	63.966	64.024	64.081	64.139	64.197	64.255	64.313	64.370
1110	64.370	64.428	64.486	64.544	64.602	64.659	64.717	64.775	64.833	64.890	64.948
1120	64.948	65.006	65.064	65.121	65.179	65.237	65.295	65.352	65.410	65.468	65.525
1130	65.525	65.583	65.641	65.699	65.756	65.814	65.872	65.929	65.987	66.045	66.102
1140	66.102	66.160	66.218	66.275	66.333	66.391	66.448	66.506	66.564	66.621	66.679

...ITS-90 Table for Type J Thermocouple

Thermoelectric Voltage in mV

°C	0	1	2	3	4	5	6	7	8	9	10
1150	66.679	66.737	66.794	66.852	66.910	66.967	67.025	67.082	67.140	67.198	67.255
1160	67.255	67.313	67.370	67.428	67.486	67.543	67.601	67.658	67.716	67.773	67.831
1170	67.831	67.888	67.946	68.003	68.061	68.119	68.176	68.234	68.291	68.348	68.406
1180	68.406	68.463	68.521	68.578	68.636	68.693	68.751	68.808	68.865	68.923	68.980
1190	68.980	69.037	69.095	69.152	69.209	69.267	69.324	69.381	69.439	69.496	69.553
1200	69.553										

ITS-90 Table for Type K Thermocouple...

Thermoelectric Voltage in mV

°C	0	−1	−2	−3	−4	−5	−6	−7	−8	−9	−10
−270	−6.458										
−260	−6.441	−6.444	−6.446	−6.448	−6.450	−6.452	−6.453	−6.455	−6.456	−6.457	−6.458
−250	−6.404	−6.408	−6.413	−6.417	−6.421	−6.425	−6.429	−6.432	−6.435	−6.438	−6.441
−240	−6.344	−6.351	−6.358	−6.364	−6.370	−6.377	−6.382	−6.388	−6.393	−6.399	−6.404
−230	−6.262	−6.271	−6.280	−6.289	−6.297	−6.306	−6.314	−6.322	−6.329	−6.337	−6.344
−220	−6.158	−6.170	−6.181	−6.192	−6.202	−6.213	−6.223	−6.233	−6.243	−6.252	−6.262
−210	−6.035	−6.048	−6.061	−6.074	−6.087	−6.099	−6.111	−6.123	−6.135	−6.147	−6.158
−200	−5.891	−5.907	−5.922	−5.936	−5.951	−5.965	−5.980	−5.994	−6.007	−6.021	−6.035
−190	−5.730	−5.747	−5.763	−5.780	−5.797	−5.813	−5.829	−5.845	−5.861	−5.876	−5.891
−180	−5.550	−5.569	−5.588	−5.606	−5.624	−5.642	−5.660	−5.678	−5.695	−5.713	−5.730
−170	−5.354	−5.374	−5.395	−5.415	−5.435	−5.454	−5.474	−5.493	−5.512	−5.531	−5.550
−160	−5.141	−5.163	−5.185	−5.207	−5.228	−5.250	−5.271	−5.292	−5.313	−5.333	−5.354
−150	−4.913	−4.936	−4.960	−4.983	−5.006	−5.029	−5.052	−5.074	−5.097	−5.119	−5.141
−140	−4.669	−4.694	−4.719	−4.744	−4.768	−4.793	−4.817	−4.841	−4.865	−4.889	−4.913
−130	−4.411	−4.437	−4.463	−4.490	−4.516	−4.542	−4.567	−4.593	−4.618	−4.644	−4.669
−120	−4.138	−4.166	−4.194	−4.221	−4.249	−4.276	−4.303	−4.330	−4.357	−4.384	−4.411
−110	−3.852	−3.882	−3.911	−3.939	−3.968	−3.997	−4.025	−4.054	−4.082	−4.110	−4.138
−100	−3.554	−3.584	−3.614	−3.645	−3.675	−3.705	−3.734	−3.764	−3.794	−3.823	−3.852
−90	−3.243	−3.274	−3.306	−3.337	−3.368	−3.400	−3.431	−3.462	−3.492	−3.523	−3.554
−80	−2.920	−2.953	−2.986	−3.018	−3.050	−3.083	−3.115	−3.147	−3.179	−3.211	−3.243
−70	−2.587	−2.620	−2.654	−2.688	−2.721	−2.755	−2.788	−2.821	−2.854	−2.887	−2.920
−60	−2.243	−2.278	−2.312	−2.347	−2.382	−2.416	−2.450	−2.485	−2.519	−2.553	−2.587
−50	−1.889	−1.925	−1.961	−1.996	−2.032	−2.067	−2.103	−2.138	−2.173	−2.208	−2.243
−40	−1.527	−1.564	−1.600	−1.637	−1.673	−1.709	−1.745	−1.782	−1.818	−1.854	−1.889
−30	−1.156	−1.194	−1.231	−1.268	−1.305	−1.343	−1.380	−1.417	−1.453	−1.490	−1.527
−20	−0.778	−0.816	−0.854	−0.892	−0.930	−0.968	−1.006	−1.043	−1.081	−1.119	−1.156
−10	−0.392	−0.431	−0.470	−0.508	−0.547	−0.586	−0.624	−0.663	−0.701	−0.739	−0.778
0	0.000	−0.039	−0.079	−0.118	−0.157	−0.197	−0.236	−0.275	−0.314	−0.353	−0.392

°C	0	1	2	3	4	5	6	7	8	9	10
0	0.000	0.039	0.079	0.119	0.158	0.198	0.238	0.277	0.317	0.357	0.397
10	0.397	0.437	0.477	0.517	0.557	0.597	0.637	0.677	0.718	0.758	0.798
20	0.798	0.838	0.879	0.919	0.960	1.000	1.041	1.081	1.122	1.163	1.203
30	1.203	1.244	1.285	1.326	1.366	1.407	1.448	1.489	1.530	1.571	1.612
40	1.612	1.653	1.694	1.735	1.776	1.817	1.858	1.899	1.941	1.982	2.023

...ITS-90 Table for Type K Thermocouple...

Thermoelectric Voltage in mV

°C	0	1	2	3	4	5	6	7	8	9	10
50	2.023	2.064	2.106	2.147	2.188	2.230	2.271	2.312	2.354	2.395	2.436
60	2.436	2.478	2.519	2.561	2.602	2.644	2.685	2.727	2.768	2.810	2.851
70	2.851	2.893	2.934	2.976	3.017	3.059	3.100	3.142	3.184	3.225	3.267
80	3.267	3.308	3.350	3.391	3.433	3.474	3.516	3.557	3.599	3.640	3.682
90	3.682	3.723	3.765	3.806	3.848	3.889	3.931	3.972	4.013	4.055	4.096
100	4.096	4.138	4.179	4.220	4.262	4.303	4.344	4.385	4.427	4.468	4.509
110	4.509	4.550	4.591	4.633	4.674	4.715	4.756	4.797	4.838	4.879	4.920
120	4.920	4.961	5.002	5.043	5.084	5.124	5.165	5.206	5.247	5.288	5.328
130	5.328	5.369	5.410	5.450	5.491	5.532	5.572	5.613	5.653	5.694	5.735
140	5.735	5.775	5.815	5.856	5.896	5.937	5.977	6.017	6.058	6.098	6.138
150	6.138	6.179	6.219	6.259	6.299	6.339	6.380	6.420	6.460	6.500	6.540
160	6.540	6.580	6.620	6.660	6.701	6.741	6.781	6.821	6.861	6.901	6.941
170	6.941	6.981	7.021	7.060	7.100	7.140	7.180	7.220	7.260	7.300	7.340
180	7.340	7.380	7.420	7.460	7.500	7.540	7.579	7.619	7.659	7.699	7.739
190	7.739	7.779	7.819	7.859	7.899	7.939	7.979	8.019	8.059	8.099	8.138
200	8.138	8.178	8.218	8.258	8.298	8.338	8.378	8.418	8.458	8.499	8.539
210	8.539	8.579	8.619	8.659	8.699	8.739	8.779	8.819	8.860	8.900	8.940
220	8.940	8.980	9.020	9.061	9.101	9.141	9.181	9.222	9.262	9.302	9.343
230	9.343	9.383	9.423	9.464	9.504	9.545	9.585	9.626	9.666	9.707	9.747
240	9.747	9.788	9.828	9.869	9.909	9.950	9.991	10.031	10.072	10.113	10.153
250	10.153	10.194	10.235	10.276	10.316	10.357	10.398	10.439	10.480	10.520	10.561
260	10.561	10.602	10.643	10.684	10.725	10.766	10.807	10.848	10.889	10.930	10.971
270	10.971	11.012	11.053	11.094	11.135	11.176	11.217	11.259	11.300	11.341	11.382
280	11.382	11.423	11.465	11.506	11.547	11.588	11.630	11.671	11.712	11.753	11.795
290	11.795	11.836	11.877	11.919	11.960	12.001	12.043	12.084	12.126	12.167	12.209
300	12.209	12.250	12.291	12.333	12.374	12.416	12.457	12.499	12.540	12.582	12.624
310	12.624	12.665	12.707	12.748	12.790	12.831	12.873	12.915	12.956	12.998	13.040
320	13.040	13.081	13.123	13.165	13.206	13.248	13.290	13.331	13.373	13.415	13.457
330	13.457	13.498	13.540	13.582	13.624	13.665	13.707	13.749	13.791	13.833	13.874
340	13.874	13.916	13.958	14.000	14.042	14.084	14.126	14.167	14.209	14.251	14.293
350	14.293	14.335	14.377	14.419	14.461	14.503	14.545	14.587	14.629	14.671	14.713
360	14.713	14.755	14.797	14.839	14.881	14.923	14.965	15.007	15.049	15.091	15.133
370	15.133	15.175	15.217	15.259	15.301	15.343	15.385	15.427	15.469	15.511	15.554
380	15.554	15.596	15.638	15.680	15.722	15.764	15.806	15.849	15.891	15.933	15.975
390	15.975	16.017	16.059	16.102	16.144	16.186	16.228	16.270	16.313	16.355	16.397

...ITS-90 Table for Type K Thermocouple...

Thermoelectric Voltage in mV

°C	0	1	2	3	4	5	6	7	8	9	10
400	16.397	16.439	16.482	16.524	16.566	16.608	16.651	16.693	16.735	16.778	16.820
410	16.820	16.862	16.904	16.947	16.989	17.031	17.074	17.116	17.158	17.201	17.243
420	17.243	17.285	17.328	17.370	17.413	17.455	17.497	17.540	17.582	17.624	17.667
430	17.667	17.709	17.752	17.794	17.837	17.879	17.921	17.964	18.006	18.049	18.091
440	18.091	18.134	18.176	18.218	18.261	18.303	18.346	18.388	18.431	18.473	18.516
450	18.516	18.558	18.601	18.643	18.686	18.728	18.771	18.813	18.856	18.898	18.941
460	18.941	18.983	19.026	19.068	19.111	19.154	19.196	19.239	19.281	19.324	19.366
470	19.366	19.409	19.451	19.494	19.537	19.579	19.622	19.664	19.707	19.750	19.792
480	19.792	19.835	19.877	19.920	19.962	20.005	20.048	20.090	20.133	20.175	20.218
490	20.218	20.261	20.303	20.346	20.389	20.431	20.474	20.516	20.559	20.602	20.644
500	20.644	20.687	20.730	20.772	20.815	20.857	20.900	20.943	20.985	21.028	21.071
510	21.071	21.113	21.156	21.199	21.241	21.284	21.326	21.369	21.412	21.454	21.497
520	21.497	21.540	21.582	21.625	21.668	21.710	21.753	21.796	21.838	21.881	21.924
530	21.924	21.966	22.009	22.052	22.094	22.137	22.179	22.222	22.265	22.307	22.350
540	22.350	22.393	22.435	22.478	22.521	22.563	22.606	22.649	22.691	22.734	22.776
550	22.776	22.819	22.862	22.904	22.947	22.990	23.032	23.075	23.117	23.160	23.203
560	23.203	23.245	23.288	23.331	23.373	23.416	23.458	23.501	23.544	23.586	23.629
570	23.629	23.671	23.714	23.757	23.799	23.842	23.884	23.927	23.970	24.012	24.055
580	24.055	24.097	24.140	24.182	24.225	24.267	24.310	24.353	24.395	24.438	24.480
590	24.480	24.523	24.565	24.608	24.650	24.693	24.735	24.778	24.820	24.863	24.905
600	24.905	24.948	24.990	25.033	25.075	25.118	25.160	25.203	25.245	25.288	25.330
610	25.330	25.373	25.415	25.458	25.500	25.543	25.585	25.627	25.670	25.712	25.755
620	25.755	25.797	25.840	25.882	25.924	25.967	26.009	26.052	26.094	26.136	26.179
630	26.179	26.221	26.263	26.306	26.348	26.390	26.433	26.475	26.517	26.560	26.602
640	26.602	26.644	26.687	26.729	26.771	26.814	26.856	26.898	26.940	26.983	27.025
650	27.025	27.067	27.109	27.152	27.194	27.236	27.278	27.320	27.363	27.405	27.447
660	27.447	27.489	27.531	27.574	27.616	27.658	27.700	27.742	27.784	27.826	27.869
670	27.869	27.911	27.953	27.995	28.037	28.079	28.121	28.163	28.205	28.247	28.289
680	28.289	28.332	28.374	28.416	28.458	28.500	28.542	28.584	28.626	28.668	28.710
690	28.710	28.752	28.794	28.835	28.877	28.919	28.961	29.003	29.045	29.087	29.129
700	29.129	29.171	29.213	29.255	29.297	29.338	29.380	29.422	29.464	29.506	29.548
710	29.548	29.589	29.631	29.673	29.715	29.757	29.798	29.840	29.882	29.924	29.965
720	29.965	30.007	30.049	30.090	30.132	30.174	30.216	30.257	30.299	30.341	30.382
730	30.382	30.424	30.466	30.507	30.549	30.590	30.632	30.674	30.715	30.757	30.798
740	30.798	30.840	30.881	30.923	30.964	31.006	31.047	31.089	31.130	31.172	31.213

...ITS-90 Table for Type K Thermocouple...

Thermoelectric Voltage in mV

°C	0	−1	−2	−3	−4	−5	−6	−7	−8	−9	−10
750	31.213	31.255	31.296	31.338	31.379	31.421	31.462	31.504	31.545	31.586	31.628
760	31.628	31.669	31.710	31.752	31.793	31.834	31.876	31.917	31.958	32.000	32.041
770	32.041	32.082	32.124	32.165	32.206	32.247	32.289	32.330	32.371	32.412	32.453
780	32.453	32.495	32.536	32.577	32.618	32.659	32.700	32.742	32.783	32.824	32.865
790	32.865	32.906	32.947	32.988	33.029	33.070	33.111	33.152	33.193	33.234	33.275
800	33.275	33.316	33.357	33.398	33.439	33.480	33.521	33.562	33.603	33.644	33.685
810	33.685	33.726	33.767	33.808	33.848	33.889	33.930	33.971	34.012	34.053	34.093
820	34.093	34.134	34.175	34.216	34.257	34.297	34.338	34.379	34.420	34.460	34.501
830	34.501	34.542	34.582	34.623	34.664	34.704	34.745	34.786	34.826	34.867	34.908
840	34.908	34.948	34.989	35.029	35.070	35.110	35.151	35.192	35.232	35.273	35.313
850	35.313	35.354	35.394	35.435	35.475	35.516	35.556	35.596	35.637	35.677	35.718
860	35.718	35.758	35.798	35.839	35.879	35.920	35.960	36.000	36.041	36.081	36.121
870	36.121	36.162	36.202	36.242	36.282	36.323	36.363	36.403	36.443	36.484	36.524
880	36.524	36.564	36.604	36.644	36.685	36.725	36.765	36.805	36.845	36.885	36.925
890	36.925	36.965	37.006	37.046	37.086	37.126	37.166	37.206	37.246	37.286	37.326
900	37.326	37.366	37.406	37.446	37.486	37.526	37.566	37.606	37.646	37.686	37.725
910	37.725	37.765	37.805	37.845	37.885	37.925	37.965	38.005	38.044	38.084	38.124
920	38.124	38.164	38.204	38.243	38.283	38.323	38.363	38.402	38.442	38.482	38.522
930	38.522	38.561	38.601	38.641	38.680	38.720	38.760	38.799	38.839	38.878	38.918
940	38.918	38.958	38.997	39.037	39.076	39.116	39.155	39.195	39.235	39.274	39.314
950	39.314	39.353	39.393	39.432	39.471	39.511	39.550	39.590	39.629	39.669	39.708
960	39.708	39.747	39.787	39.826	39.866	39.905	39.944	39.984	40.023	40.062	40.101
970	40.101	40.141	40.180	40.219	40.259	40.298	40.337	40.376	40.415	40.455	40.494
980	40.494	40.533	40.572	40.611	40.651	40.690	40.729	40.768	40.807	40.846	40.885
990	40.885	40.924	40.963	41.002	41.042	41.081	41.120	41.159	41.198	41.237	41.276
1000	41.276	41.315	41.354	41.393	41.431	41.470	41.509	41.548	41.587	41.626	41.665
1010	41.665	41.704	41.743	41.781	41.820	41.859	41.898	41.937	41.976	42.014	42.053
1020	42.053	42.092	42.131	42.169	42.208	42.247	42.286	42.324	42.363	42.402	42.440
1030	42.440	42.479	42.518	42.556	42.595	42.633	42.672	42.711	42.749	42.788	42.826
1040	42.826	42.865	42.903	42.942	42.980	43.019	43.057	43.096	43.134	43.173	43.211
1050	43.211	43.250	43.288	43.327	43.365	43.403	43.442	43.480	43.518	43.557	43.595
1060	43.595	43.633	43.672	43.710	43.748	43.787	43.825	43.863	43.901	43.940	43.978
1070	43.978	44.016	44.054	44.092	44.130	44.169	44.207	44.245	44.283	44.321	44.359
1080	44.359	44.397	44.435	44.473	44.512	44.550	44.588	44.626	44.664	44.702	44.740
1090	44.740	44.778	44.816	44.853	44.891	44.929	44.967	45.005	45.043	45.081	45.119

...ITS-90 Table for Type K Thermocouple

Thermoelectric Voltage in mV

°C	0	−1	−2	−3	−4	−5	−6	−7	−8	−9	−10
1100	45.119	45.157	45.194	45.232	45.270	45.308	45.346	45.383	45.421	45.459	45.497
1110	45.497	45.534	45.572	45.610	45.647	45.685	45.723	45.760	45.798	45.836	45.873
1120	45.873	45.911	45.948	45.986	46.024	46.061	46.099	46.136	46.174	46.211	46.249
1130	46.249	46.286	46.324	46.361	46.398	46.436	46.473	46.511	46.548	46.585	46.623
1140	46.623	46.660	46.697	46.735	46.772	46.809	46.847	46.884	46.921	46.958	46.995
1150	46.995	47.033	47.070	47.107	47.144	47.181	47.218	47.256	47.293	47.330	47.367
1160	47.367	47.404	47.441	47.478	47.515	47.552	47.589	47.626	47.663	47.700	47.737
1170	47.737	47.774	47.811	47.848	47.884	47.921	47.958	47.995	48.032	48.069	48.105
1180	48.105	48.142	48.179	48.216	48.252	48.289	48.326	48.363	48.399	48.436	48.473
1190	48.473	48.509	48.546	48.582	48.619	48.656	48.692	48.729	48.765	48.802	48.838
1200	48.838	48.875	48.911	48.948	48.984	49.021	49.057	49.093	49.130	49.166	49.202
1210	49.202	49.239	49.275	49.311	49.348	49.384	49.420	49.456	49.493	49.529	49.565
1220	49.565	49.601	49.637	49.674	49.710	49.746	49.782	49.818	49.854	49.890	49.926
1230	49.926	49.962	49.998	50.034	50.070	50.106	50.142	50.178	50.214	50.250	50.286
1240	50.286	50.322	50.358	50.393	50.429	50.465	50.501	50.537	50.572	50.608	50.644
1250	50.644	50.680	50.715	50.751	50.787	50.822	50.858	50.894	50.929	50.965	51.000
1260	51.000	51.036	51.071	51.107	51.142	51.178	51.213	51.249	51.284	51.320	51.355
1270	51.355	51.391	51.426	51.461	51.497	51.532	51.567	51.603	51.638	51.673	51.708
1280	51.708	51.744	51.779	51.814	51.849	51.885	51.920	51.955	51.990	52.025	52.060
1290	52.060	52.095	52.130	52.165	52.200	52.235	52.270	52.305	52.340	52.375	52.410
1300	52.410	52.445	52.480	52.515	52.550	52.585	52.620	52.654	52.689	52.724	52.759
1310	52.759	52.794	52.828	52.863	52.898	52.932	52.967	53.002	53.037	53.071	53.106
1320	53.106	53.140	53.175	53.210	53.244	53.279	53.313	53.348	53.382	53.417	53.451
1330	53.451	53.486	53.520	53.555	53.589	53.623	53.658	53.692	53.727	53.761	53.795
1340	53.795	53.830	53.864	53.898	53.932	53.967	54.001	54.035	54.069	54.104	54.138
1350	54.138	54.172	54.206	54.240	54.274	54.308	54.343	54.377	54.411	54.445	54.479
1360	54.479	54.513	54.547	54.581	54.615	54.649	54.683	54.717	54.751	54.785	54.819
1370	54.819	54.852	54.886								

Psychrometric Chart

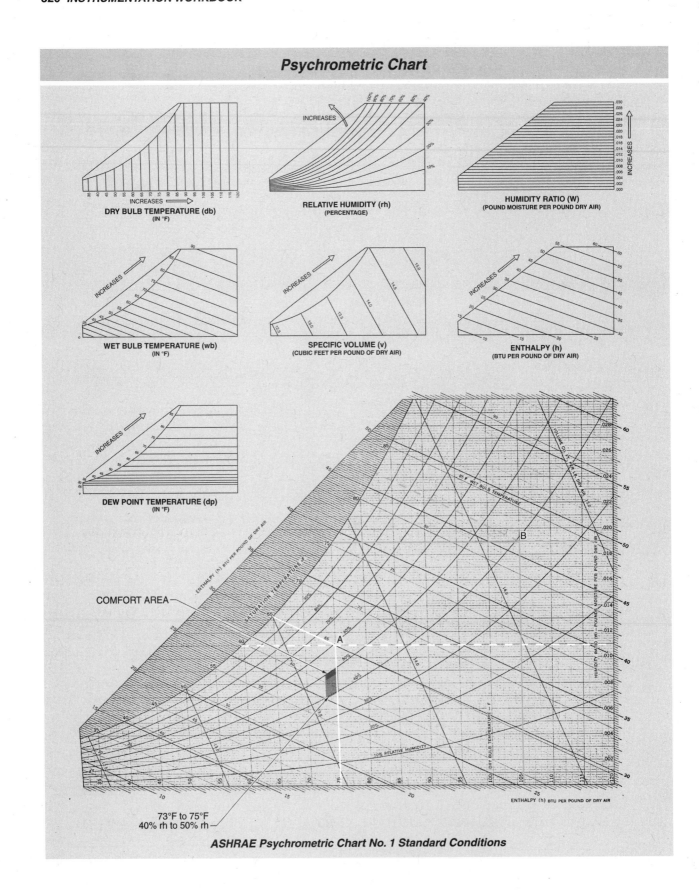

ASHRAE Psychrometric Chart No. 1 Standard Conditions

Dry Saturated Steam Pressure...

Absolute Pressure* p	Temperature† t	Specific Volume‡		Enthalpy‖			Entropy		
		Saturated Liquid, v_f	Saturated Vapor, v_g	Saturated Liquid, h_f	Evap. h_{fg}	Saturated Vapor, h_g	Saturated Liquid, s_f	Evap. s_{fg}	Saturated Vapor, s_g
1	101.74	0.01614	333.6	69.70	1036.3	1106.0	0.1326	1.8456	1.9782
2	126.08	0.01623	173.73	93.99	1022.2	1116.2	0.1749	1.7451	1.9200
3	141.48	0.01630	118.71	109.37	1013.2	1122.6	0.2008	1.6855	1.8863
4	152.97	0.01636	90.63	120.86	1006.4	1127.3	0.2198	1.6427	1.8625
5	164.24	0.01640	73.52	130.13	1001.0	1131.1	0.2347	1.6094	1.8441
6	170.06	0.01645	61.98	137.96	996.2	1134.2	0.2472	1.5820	1.8292
7	176.85	0.01649	53.64	144.76	992.1	1136.9	0.2581	1.5586	1.8167
8	182.86	0.01653	47.34	150.79	989.5	1139.3	0.2674	1.5383	1.8057
9	188.28	0.01656	42.40	156.22	985.2	1141.4	0.2759	1.5203	1.7962
10	193.21	0.01659	38.42	161.17	982.1	1143.3	0.2835	1.5041	1.7876
14.696	212.00	0.01672	26.80	180.07	970.3	1150.4	0.3120	1.4446	1.7566
15	213.03	0.01684	26.29	181.11	969.7	1150.8	0.3135	1.4415	1.7549
20	227.96	0.01683	20.089	196.16	960.1	1156.3	0.3356	1.3962	1.7319
25	240.07	0.01692	16.303	208.42	952.1	1160.6	0.3533	1.3606	1.7139
30	250.33	0.01701	13.746	218.82	945.3	1164.1	0.3680	1.3313	1.6993
35	259.28	0.01708	11.898	227.91	939.2	1167.1	0.3807	1.3063	1.6870
40	267.25	0.01715	10.498	236.06	933.7	1169.7	03919	1.2844	1.6763
45	274.44	0.01721	9.401	243.36	928.6	1172.0	0.4019	1.2650	1.6669
50	281.01	0.01727	8.515	250.09	924.0	1174.1	0.4110	1.2474	1.6585
55	287.07	0.01732	7.787	256.30	919.6	1175.9	0.4193	1.2316	1.6509
60	292.71	0.01738	7.175	262.09	915.5	1177.6	0.4270	1.2168	1.6438
65	297.97	0.01743	6.655	267.50	911.6	1179.1	0.4342	1.2032	1.6374
70	302.92	0.01748	6.203	272.61	907.9	1180.3	0.4409	1.1906	1.6315
75	307.60	0.01753	5.816	277.43	904.5	1181.9	0.4472	1.1787	1.6259
80	312.03	0.01757	5.472	282.02	901.1	1183.1	0.4531	1.1676	1.6207
85	316.25	0.01761	5.168	286.39	897.8	1184.2	0.4587	1.1571	1.6158
90	320.27	0.01766	4.896	290.56	894.7	1185.3	0.4641	1.1471	1.6112
95	324.12	0.01770	4.652	294.56	891.7	1186.2	0.4692	1.1376	1.6068
100	327.81	0.01774	4.432	298.40	888.8	1187.2	0.4740	1.1286	1.6026
110	334.77	0.01782	4.049	305.66	883.2	1188.9	0.4832	1.1117	1.5948
120	341.25	0.01789	3.72	312.44	877.9	1190.4	0.4916	1.0962	1.5878
130	347.32	0.01796	3.455	318.81	872.9	1191.7	0.4995	1.0817	1.5812
140	353.02	0.01802	3.220	324.82	868.2	1193.0	0.5069	1.0682	1.5751
150	358.42	0.01809	3.015	330.51	863.6	1194.1	0.5138	1.0556	1.5694
160	363.53	0.01815	2.834	335.93	859.2	1195.1	0.5204	1.0436	1.5640
170	368.41	0.01822	2.675	341.09	854.9	1196.0	0.5266	1.0324	1.5590
180	373.06	0.01827	2.532	346.03	850.8	1196.9	0.5325	1.0217	1.5542
190	377.51	0.01833	2.404	350.79	846.8	1197.6	0.5381	1.0116	1.5497
200	381.79	0.01839	2.288	355.36	843.0	1198.4	0.5435	1.0018	1.5453
250	400.95	0.01865	1.8438	376.00	825.1	1201.1	0.5675	0.9588	1.5263
300	417.33	0.01890	1.5433	393.84	809.0	1202.8	0.5879	0.9225	1.5104

* in psi ‡ in cu ft/lb

† in °F ‖ in Btu/lb

...Dry Saturated Steam Pressure

Absolute Pressure* p	Temperature† t	Specific Volume‡		Enthalpy‖			Entropy		
		Saturated Liquid, v_f	Saturated Vapor, v_g	Saturated Liquid, h_f	Evap. h_{fg}	Saturated Vapor, h_g	Saturated Liquid, s_f	Evap. s_{fg}	Saturated Vapor, s_g
350	431.72	0.01913	1.3260	409.69	794.2	1203.9	0.6056	.8910	1.4966
400	444.59	0.0193	1.1613	424.0	780.5	1204.5	0.6214	.8630	1.4844
450	456.28	0.0195	1.0320	437.2	767.4	1204.6	0.6356	.8378	1.4734
500	467.01	0.0197	0.9278	449.4	755.0	1204.4	0.6487	.8147	1.4634
550	476.94	0.0199	0.8424	460.8	743.1	1203.9	0.6608	.7934	1.4542
600	486.21	0.0201	0.7698	471.6	731.6	1203.2	0.6720	.7734	1.4454
650	494.90	0.0203	0.7083	481.8	720.5	1202.3	0.6826	.7548	1.4374
700	503.10	0.0205	0.6554	491.5	709.7	1201.2	0.6925	.7371	1.4296
750	510.86	0.0207	0.6092	500.8	699.2	1200.0	0.7019	.7204	1.4223
800	518.23	0.0209	0.5687	509.7	688.9	1198.6	0.7108	.7045	1.4153
850	525.26	0.0210	0.5327	518.3	678.8	1197.7	0.7194	.6891	1.4085
900	531.98	0.0212	0.5006	526.6	668.8	1195.4	0.7275	.6744	1.4020
950	538.43	0.0214	0.4717	534.6	659.1	1193.7	0.7355	.6602	1.3957
1000	544.61	0.0216	0.4456	542.4	649.4	1191.8	0.7430	.6467	1.3897
1100	556.31	0.0220	0.4001	557.4	630.4	1187.8	0.7575	.6205	1.3780
1200	567.22	0.0223	0.3619	571.7	611.7	1183.4	0.7711	.5956	1.3667
1300	577.46	0.0227	0.3293	585.4	593.2	1178.6	0.7840	.5719	1.3559
1400	587.10	0.0231	0.3012	598.7	574.7	1173.4	0.7963	.5491	1.3454
1500	596.23	0.0235	0.2765	611.6	556.3	1167.9	0.8082	.5269	1.3351
2000	635.82	0.0257	0.1878	671.7	463.4	1135.1	0.8619	.4230	1.2849
2500	668.13	0.0287	0.1307	730.6	360.5	1091.1	0.9126	.3197	1.2322
3000	695.36	0.0346	0.0858	802.5	217.8	1020.3	0.9731	.1885	1.1615
3206.2	705.40	0.0503	0.0503	902.7	0	902.7	1.0580	0	1.0580

* in psi ‡ in cu ft/lb
† in °F ‖ in Btu/lb

Instrument Tag Identification

	First Letter		Second Letter		
	Measured or Initiating Variable	**Modifier**	**Readout or Passive Function**	**Output Function**	**Modifier**
A	Analysis		Alarm		
B	Burner Flame		User's Choice	User's Choice	User's Choice
C	Conductivity (Electrical)			Control	
D	Density (Mass) or Specific Gravity	Differential			
E	Voltage (EMF)		Primary Element		
F	Flow Rate	Ratio (Fraction)			
G	Gaging (Dimensional)		Glass		
H	Hand (Manually Initiated)				High
I	Current (Electrical)		Indicate		
J	Power	Scan			
K	Time or Time Schedule			Control Station	
L	Level		Light (Pilot)		Low
M	Moisture or Humidity				Middle or Intermediate
N	User's Choice		User's Choice	User's Choice	User's Choice
O	User's Choice		Orifice (Restriction)		
P	Pressure or Vacuum		Point (Test Connection)		
Q	Quantity or Event	Integrate or Totalize			
R	Radioactivity, radiation		Record or Print		
S	Speed or Frequency	Safety		Switch	
T	Temperature			Transmit	
U	Multivariable		Multifunction	Multifunction	Multifunction
V	Viscosity, Vibration			Valve, Damper, or Louver	
W	Weight or Force		Well		
X	Unclassified		Unclassified	Unclassified	Unclassified
Y	Event or State			Relay or Compute	
Z	Position			Drive, Actuate, or Unclassified Final Control Element	

Selected Instrumentation Symbols

General Instrument Symbols — Balloons

Instrument for single measured variable * with any number of functions.

Instrument for two measured variables.* Optionally, single-variable instrument with more than one function. Additional tangent balloons my be added as required.

APPROXIMATELY ⁷⁄₁₆″ DIAMETER

LOCALLY MOUNTED

MOUNTED ON BOARD 1 (OR BOARD 2). BOARD 2 MAY ALTERNATIVELY BE DESIGNATED BY A DOUBLE HORIZONTAL LINE INSTEAD OF A SINGLE LINE, WITH THE DESIGNATION OUTSIDE THE BALLOON OMITTED.

MOUNTED BEHIND THE BOARD

LOCALLY MOUNTED INSTRUMENT WITH LONG TAG NUMBER. (6 IS OPTIONAL AND IS PLANT NUMBER.) ALTERNATIVELY, A CLOSED CIRCLE MAY BE ENLARGED.

LOCALLY MOUNTED

MOUNTED ON MAIN BOARD

Control Valve Body Symbols

GLOBE, GATE, OR OTHER IN-LINE TYPE NOT OTHERWISE IDENTIFIED

ANGLE

BUTTERFLY, DAMPER, OR LOUVER

ROTARY PLUG OR BALL

THREE-WAY

FOUR-WAY

Actuator Symbols

Diaphragm, spring-opposed

Diaphragm, spring-opposed, with positioner and overriding pilot valve that pressurizes diaphragm when actuated.

WITHOUT POSITIONER OR OTHER PILOT

PREFERRED FOR DIAPHRAGM THAT IS ASSEMBLED WITH PILOT SO THAT ASSEMBLY IS ACTUATED BY ONE CON-TROLLED INPUT (SHOWN TYPICALLY WITH ELECTRIC INPUT TO ASSEMBLY)

AIR SUPPLY

PREFERRED ALTERNATIVE

AIR SUPPLY

OPTIONAL ALTERNATIVE

DIAPHRAGM, PRESSURE-BALANCED

Cylinder, without positioner or other pilot

SINGLE-ACTING

DOUBLE-ACTING

PREFERRED FOR ANY CYLINDER THAT IS ASSEMBLED WITH PILOT SO THAT ASSEMBLY IS ACTUATED BY ONE CONTROLLED INPUT

ROTARY MOTOR (SHOWN TYPICALLY WITH ELECTRIC SIGNAL)

Symbols for Self-Actuated Regulators, Valves, and Other Devices

PCV 17 — PRESSURE-REDUCING REGULATOR, SELF-CONTAINED

PCV 18 — PRESSURE-REDUCING REGULATOR WITH EXTERNAL PRESSURE TAP

PDCV 19 — DIFFERENTIAL-PRESSURE-REDUCING REGULATOR WITH INTERNAL AND EXTERNAL PRESSURE TAPS

PCV 20 — BACKPRESSURE REGULATOR, SELF-CONTAINED

PCV 21 — BACKPRESSURE REGULATOR WITH EXTERNAL PRESSURE TAP

Symbols for Actuator Action in Event of Actuator Power Failure

FO — TWO-WAY VALVE, FAIL OPEN

FC — TWO-WAY VALVE, FAIL CLOSED

A B C FO — THREE-WAY VALVE, FAIL OPEN TO PATH A-C

FO A B C D FO — FOUR-WAY VALVE, FAIL OPEN TO PATHS A-C AND D-B

FL — ANY VALVE, FAIL LOCKED (POSITION DOES NOT CHANGE)

Selected Primary Element Symbols...

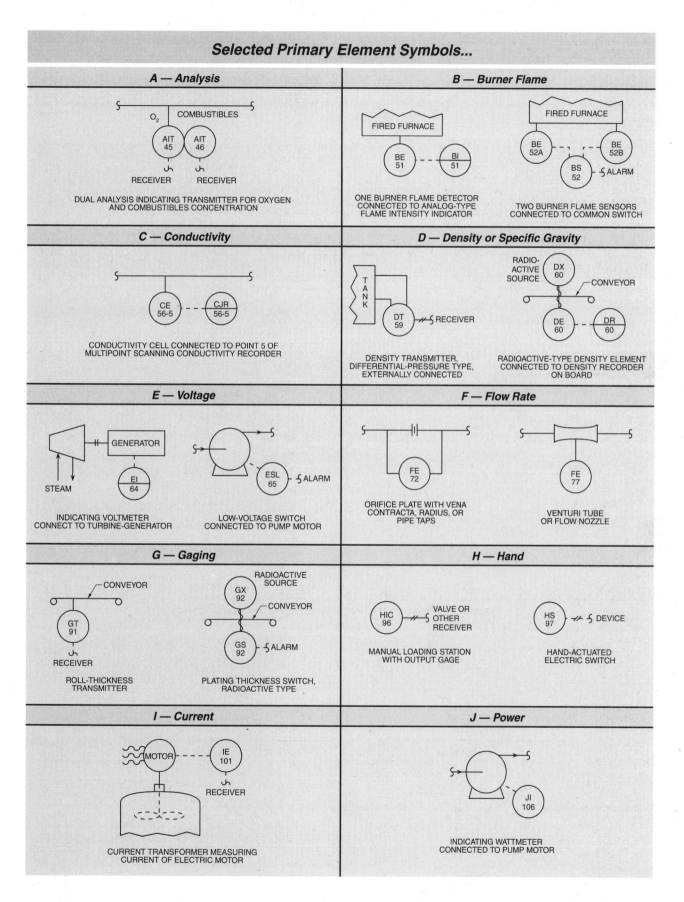

A — Analysis

DUAL ANALYSIS INDICATING TRANSMITTER FOR OXYGEN AND COMBUSTIBLES CONCENTRATION

B — Burner Flame

ONE BURNER FLAME DETECTOR CONNECTED TO ANALOG-TYPE FLAME INTENSITY INDICATOR

TWO BURNER FLAME SENSORS CONNECTED TO COMMON SWITCH

C — Conductivity

CONDUCTIVITY CELL CONNECTED TO POINT 5 OF MULTIPOINT SCANNING CONDUCTIVITY RECORDER

D — Density or Specific Gravity

DENSITY TRANSMITTER, DIFFERENTIAL-PRESSURE TYPE, EXTERNALLY CONNECTED

RADIOACTIVE-TYPE DENSITY ELEMENT CONNECTED TO DENSITY RECORDER ON BOARD

E — Voltage

INDICATING VOLTMETER CONNECT TO TURBINE-GENERATOR

LOW-VOLTAGE SWITCH CONNECTED TO PUMP MOTOR

F — Flow Rate

ORIFICE PLATE WITH VENA CONTRACTA, RADIUS, OR PIPE TAPS

VENTURI TUBE OR FLOW NOZZLE

G — Gaging

ROLL-THICKNESS TRANSMITTER

PLATING THICKNESS SWITCH, RADIOACTIVE TYPE

H — Hand

MANUAL LOADING STATION WITH OUTPUT GAGE

HAND-ACTUATED ELECTRIC SWITCH

I — Current

CURRENT TRANSFORMER MEASURING CURRENT OF ELECTRIC MOTOR

J — Power

INDICATING WATTMETER CONNECTED TO PUMP MOTOR

...Selected Primary Element Symbols

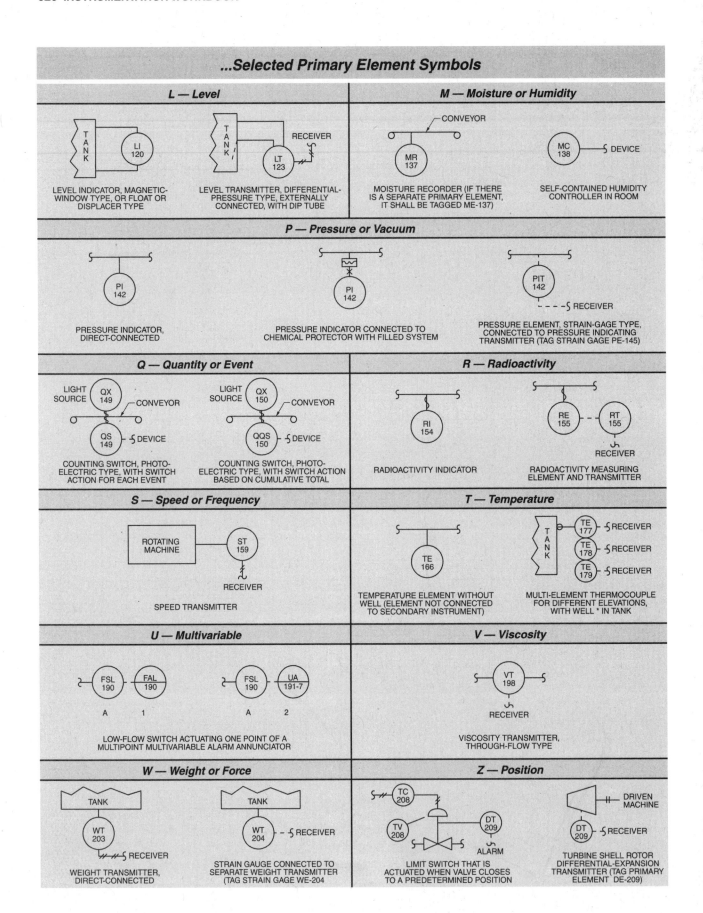

L — Level

LI 120

LEVEL INDICATOR, MAGNETIC-WINDOW TYPE, OR FLOAT OR DISPLACER TYPE

LT 123 — RECEIVER

LEVEL TRANSMITTER, DIFFERENTIAL-PRESSURE TYPE, EXTERNALLY CONNECTED, WITH DIP TUBE

M — Moisture or Humidity

CONVEYOR

MR 137

MOISTURE RECORDER (IF THERE IS A SEPARATE PRIMARY ELEMENT, IT SHALL BE TAGGED ME-137)

MC 138 — DEVICE

SELF-CONTAINED HUMIDITY CONTROLLER IN ROOM

P — Pressure or Vacuum

PI 142

PRESSURE INDICATOR, DIRECT-CONNECTED

PI 142

PRESSURE INDICATOR CONNECTED TO CHEMICAL PROTECTOR WITH FILLED SYSTEM

PIT 142 — — — RECEIVER

PRESSURE ELEMENT, STRAIN-GAGE TYPE, CONNECTED TO PRESSURE INDICATING TRANSMITTER (TAG STRAIN GAGE PE-145)

Q — Quantity or Event

LIGHT SOURCE **QX 149** — CONVEYOR

QS 149 — DEVICE

COUNTING SWITCH, PHOTO-ELECTRIC TYPE, WITH SWITCH ACTION FOR EACH EVENT

LIGHT SOURCE **QX 150** — CONVEYOR

QQS 150 — DEVICE

COUNTING SWITCH, PHOTO-ELECTRIC TYPE, WITH SWITCH ACTION BASED ON CUMULATIVE TOTAL

R — Radioactivity

RI 154

RADIOACTIVITY INDICATOR

RE 155 — — **RT 155**

RECEIVER

RADIOACTIVITY MEASURING ELEMENT AND TRANSMITTER

S — Speed or Frequency

ROTATING MACHINE **ST 159**

RECEIVER

SPEED TRANSMITTER

T — Temperature

TE 166

TEMPERATURE ELEMENT WITHOUT WELL (ELEMENT NOT CONNECTED TO SECONDARY INSTRUMENT)

TANK **TE 177** — RECEIVER
TE 178 — RECEIVER
TE 179 — RECEIVER

MULTI-ELEMENT THERMOCOUPLE FOR DIFFERENT ELEVATIONS, WITH WELL * IN TANK

U — Multivariable

FSL 190 **FAL 190**

A — 1

FSL 190 **UA 191-7**

A — 2

LOW-FLOW SWITCH ACTUATING ONE POINT OF A MULTIPOINT MULTIVARIABLE ALARM ANNUNCIATOR

V — Viscosity

VT 198

RECEIVER

VISCOSITY TRANSMITTER, THROUGH-FLOW TYPE

W — Weight or Force

TANK

WT 203

RECEIVER

WEIGHT TRANSMITTER, DIRECT-CONNECTED

TANK

WT 204 — RECEIVER

STRAIN GAUGE CONNECTED TO SEPARATE WEIGHT TRANSMITTER (TAG STRAIN GAGE WE-204

Z — Position

TC 208

TV 208 **DT 209**

ALARM

LIMIT SWITCH THAT IS ACTUATED WHEN VALVE CLOSES TO A PREDETERMINED POSITION

DRIVEN MACHINE

DT 209 — RECEIVER

TURBINE SHELL ROTOR DIFFERENTIAL-EXPANSION TRANSMITTER (TAG PRIMARY ELEMENT DE-209)